T0011306

Published by Wooden Books LLC,
San Rafael, California

First published in the UK in 2022
by Wooden Books LTD, Glastonbury

Library of Congress Cataloging-in-Publication Data
Marshall, S.
Acoustics

Library of Congress Cataloging-in-Publication
Data has been applied for

ISBN-10: 1-952178-33-9
ISBN-13: 978-1-952178-33-7

Designed and typeset in Glastonbury, UK

Printed in China on FSC® certified papers by
RR Donnelley Asia Printing Solutions Ltd.

ACOUSTICS

THE ART OF SOUND

Steve Marshall

To Kim

Thanks to Matt Tweed at Wooden Books for editorial and graphical assistance. A companion to this volume, Harmonograph *by Anthony Ashton, 2005, covers the intricacies of tuning. Other recommended books on the subject include:* On the Sensations of Tone *by Hermann von Helmholtz, 1863;* The Science of Musical Sound, *John Robinson Pierce, 1983;* Master Handbook of Acoustics, *F Alton Everest & Ken Pohlmann, 2014.*

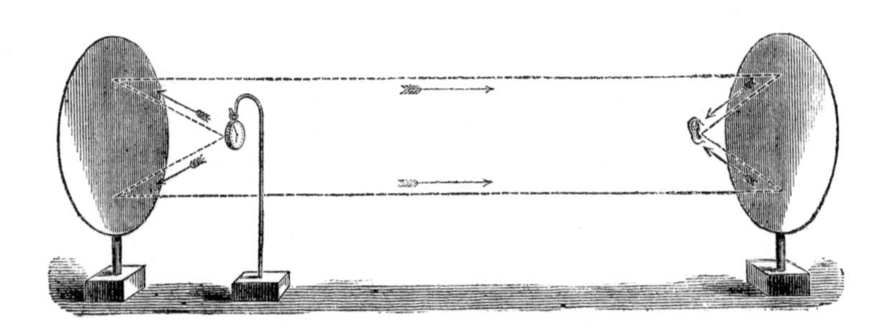

ABOVE: *A pair of parabolic reflectors can be used to focus and transmit sound. Wildlife sound recordists use parabolic reflectors for recording distant bird song. The soundwaves hit the reflector and are focused into a microphone. During World War II, parabolic reflectors 10 m in diameter were cast in concrete on the British coast, enabling the sound of approaching planes to be detected from many miles away.* TITLE PAGE: *The Whispering Gallery in St Paul's Cathedral, London, allows a whisper to travel the whole circumference as the sound waves are reflected onwards and cling to the walls.* FRONTISPIECE: *A sophisticated megaphone.*

CONTENTS

Above: *A Lecture on Acoustics. Gabriel Lippmann [1845–1921] demonstrates that vibrations can propagate through glass and along a metal rod. The resonant body of a violin is then used as an amplifier, allowing sounds from outside to be heard in the room.*

INTRODUCTION

ALL THE UNIVERSE is vibration—from atomic particles to the swirl of galaxies, from microcosm to macrocosm. What we call 'sound' is a vibration, carried by a medium such as air, water or metal, then delivered into an ear and converted and interpreted by a brain.

Many creation myths begin with a sound. The Greek philosopher Heraclitus [c.535–475 BC] introduced the concept of *Logos*, the origin of everything, personified by Hermes, inventor of the lyre. Known as *Memra* in the Jewish tradition, the *Logos* became the seed of creation in Christianity; the Gospel of St. John opens: "*In the beginning was the Word, and the Word was with God, and the Word was God*".

The Sufis of Islam believe that the intoxicating 'sound of the abstract' *Saut-e Sarmad* fills all space and is the origin and source of all mystical knowledge. Sufis train themselves to hear the Sarmad in horns, shells, bells, gongs, the roaring of the sea, the buzzing of bees and the singing of birds, until it finally evolves into the most sacred and universal creative sound of all, *Hu*, personalised in the sound *Ham*, or *Hum*.

Hindu scriptures, including the Bhagavad Gita, say that the eternal creative sound is *Om*, or more correctly *Aum*. For Taoists, this is *Kung*, the Great Tone of Nature. Buddhists chant *Om Mani padme hum*.

The North American Cherokee believe that the primary tone is sung by the quartz crystal, a vibration which governs our digital technology even today. Indeed, modern theories suggest that at the birth of the universe in the Big Bang, the initial super hot, dense ball of plasma probably rang like a vast bell, the vibrations of its birth rippling across the new cosmos to form the gravitational seeds of the galaxies.

WHAT IS SOUND?

in the beginning

SOUND IS VIBRATION, but not all vibrations are sound; we normally use the word to describe only those vibrations that can be detected by the ear, with its inherently limited range.

We 'hear' when vibrations from a sound source (such as a twanged guitar string) are transmitted to air molecules. Air is naturally elastic and its molecules vibrate in sympathy with the physical motion of the original sound source. The vibrating air molecules bounce off their neighbours and pass the movement along, through the air, until it reaches an eardrum, where it is converted into electrical nerve impulses that are sent to our brain and interpreted as sound (*see below*).

Sound waves are invisible to us, but are similar to ripples moving across a pond or the wind blowing waves across a field of barley. Although the individual components (water molecules or stalks) are fixed and only oscillate in place, we perceive an apparent movement across the entire field. Unlike these two-dimensional analogies, sound in three-dimensional and radiates in every direction.

Sound contributes to the sense of wonder and majesty we experience in sacred buildings. Many cathedrals, temples and sacred spaces around the world have extraordinary acoustics (*see examples opposite*).

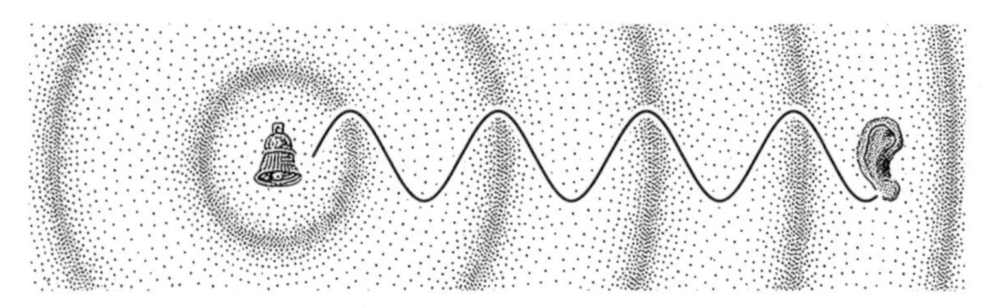

ABOVE: *Archaeoacoustic research has revealed connections between prehistoric rock art and acoustics. Some prehistoric cave paintings, like those in the Vézere Valley in France, show herds of grazing cattle. Striking rocks together in the cave produces a 'flutter echo' from the parallel walls that sounds uncannily like hoof beats.*

ABOVE LEFT: *Pulsed trumpet blasts bring down the Walls of Jericho.* ABOVE RIGHT: *The side chambers of the 3000 BC West Kennet long barrow, near Avebury, resonate at 110 Hz and 84 Hz creating a 4:3 harmony, the musical fourth. Many ancient sites resonate around 10 Hz and below. Although inaudible, infrasound can have a hypnotic effect on the human brain, producing impressions of a 'presence' and triggering altered states of consciousness and hallucinations.*

SIMPLE SOUNDS

frequencies and oscillating systems

One of the central elements of acoustics is the deceptively simple SINE WAVE. This is the wave created by the tip of a propellor as an airplane flies along, seen from the side. The sine wave emerges from circular motion. It is the pure tone we hear when a wineglass is played with a wet finger. A bouncing weight can also draw a sine wave, graphically illustrating cyclic motion plotted against time (*opposite, top right*).

The rate of vibration, or FREQUENCY, of a sound determines what we perceive as PITCH, with high pitched sounds vibrating more rapidly than low ones. Frequencies are measured in cycles per second (cps), or hertz (Hz), named for the physicist Heinrich Hertz [1857–1894]. The two are exactly the same. Like other SI Units, a prefix system is used, so 1000 hertz is 1 kilohertz, or 1kHz, and so on (*see below*).

The lower and upper limits of human hearing lie between 20 Hz and 20,000 Hz. Below that, INFRASOUND is generated by sea waves, explosions, earthquakes and geological processes. Propogating over long distances, it can reveal deep earth structures and oil deposits. At the other extreme, ULTRASOUND can harmlessly scan body tissues, clean delicate objects, find invisible faults and accurately measure distances.

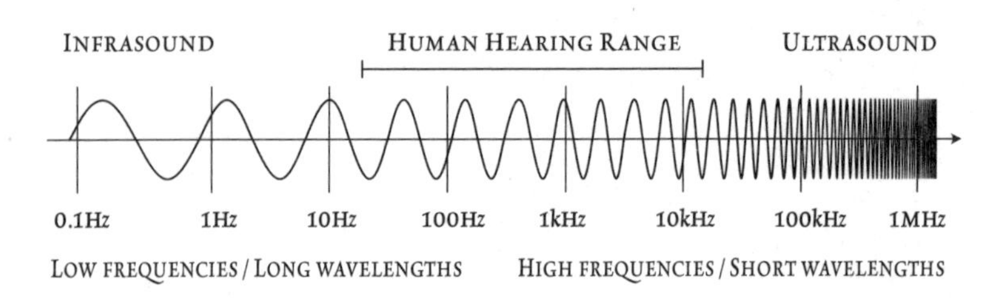

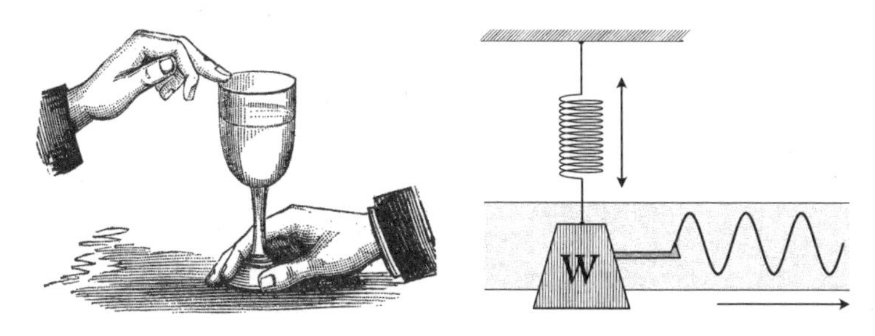

ABOVE LEFT: *Playing a wineglass with a wet finger produces the pure tone of a sine wave.*
ABOVE RIGHT: *A heavy weight suspended from a spring oscillates in simple harmonic motion. An attached pencil will then draw a sine wave on a strip of paper pulled past at constant speed.*

ABOVE: *A sine wave generated by a large tuning fork is inscribed onto a smoke-blackened cylinder. A similar device, the phonautograph, was the first to record actual sounds; it used a horn, attached to a stiff bristle via a flexible diaphragm, to transcribe soundwaves onto a sooty cylinder.*

WAVEFORMS
amplitude and wavelength

Sounds are usually invisible, but electronic tools like oscilloscopes can reveal several defining features of their waveforms. The AMPLITUDE is the size of the wave, which we perceive as the loudness of a sound. WAVELENGTH is the distance over which a wave repeats, denoted by the Greek letter *lambda* (λ). The wavelength is the inverse of frequency (*see page 4*), with the velocity of a wave being its wavelength multiplied by its frequency (*opposite top*). For sound, this relationship means $\lambda = \frac{c}{f}$ where c is the speed of sound and f is the frequency.

Each sound has a distinctive WAVEFORM. Blowing across the end of an open pipe will produce a SINE wave, whereas a stopped pipe will make a hollow-sounding SQUARE wave. Bowed instruments, like the violin, generate SAWTOOTH waves, as the sticky hairs of the bow drag the string in one direction until it snaps back; repeated many times a second, the process can be seen in the resulting waveform (*lower opposite*).

Adding together two identical waveforms with opposite PHASE results in silence (*right*). This is used by 'noise-reducing' headphones, where a microphone collects ambient sounds around the listener, and mixes these back in phase-reversed with whatever is playing through the headphones. Phase-cancellation then causes the ambient noise to disappear.

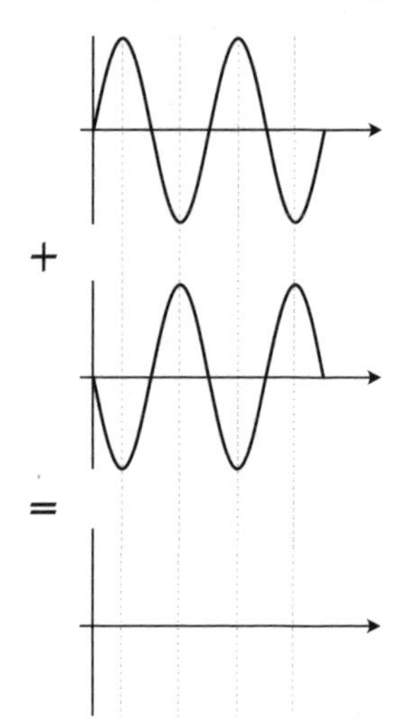

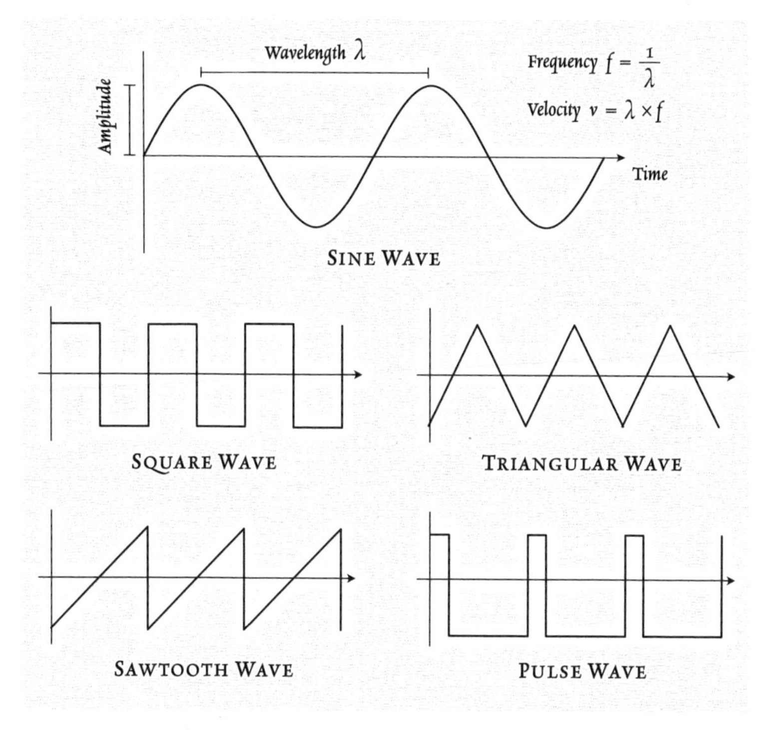

ABOVE: *Basic waveforms, showing amplitude plotted against time. Sine waves can be heard in the sound of a tuning fork or a softly blown flute. Square waves form as the fundamental tones of closed pipes. Triangular waves can be approximated by a guitar string plucked at its mid-point. Sawtooth waves are the raw sound of bowed instruments. Pulse waves are produced by rotating disc sirens.*

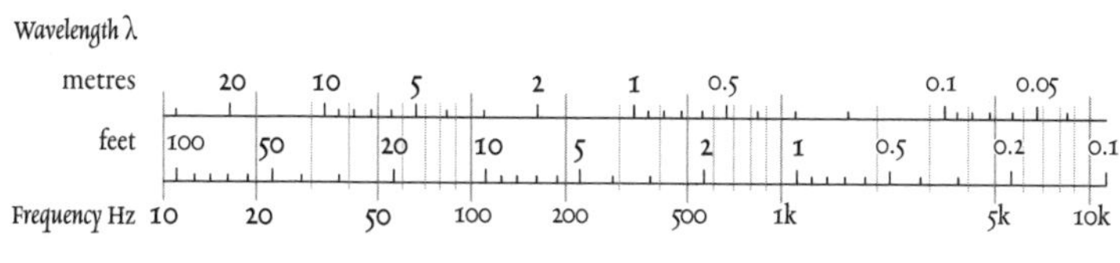

ABOVE: *Sound wavelengths at various frequencies.*

THE SPEED OF SOUND
and sonic booms

The speed of sound is not fixed, but varies depending on the medium it travels in. Sound moves faster in denser mediums—it is slow in gases, faster in liquids and fastest in solids. In dry air air at 20°C its speed is about 343 m/s, or 768 miles per hour (mph). In iron it is 4,994 m/s, or 11,170 mph. In fresh water at 20°C sound travels at 1481m/s, but in denser sea water this increases to 1521m/s. Temperature matters too—hotter means faster. In dry air at 0°C sound travels at a sluggish 331m/s, whereas in hot air at 100°C the speed of sound is 387 m/s. Higher humidity in air also speeds up sound waves slightly. The speed of sound does not vary with frequency, amplitude or wavelength.

Flying faster than the speed of sound creates a SONIC BOOM—a shock wave that can damage roofs and break windows (*lower opposite*). The bang of a gunshot is partially a sonic boom generated by the bullet moving faster than sound; similarly a whip crack is caused by the end going supersonic.

Sound travels about a million times more slowly than light, so to estimate how far away an electrical storm is, count the seconds between seeing a lightning flash and hearing the clap of thunder. Sound covers one km in 2.9 seconds, or a mile in 4.7 seconds. So, for the distance in kilometres divide the time taken in seconds by 3, for the distance in miles divide by 5.

ABOVE: *How fast is the speed of sound in air? In one experiment a cannon is fired, and an observation post a known distance away starts a clock on seeing its flash, and stops it upon hearing the bang. The speed of sound is given by dividing the distance by the time between flash and bang.*

ABOVE: *Breaking the sound barrier. Flying at the speed of sound (Mach 1) creates a 'sonic boom', a shock wave caused by sound waves catching up and overtaking each other as a plane goes supersonic. Faster still, and the roar of the jet appears to come from behind the plane.*

The Ear
from moths to whales

Sound is a construct. Our hearing system converts vibrations of the air into electrical nerve impulses that our brain interprets as 'sound'.

The soft outer ear, the PINNA (A, *opposite*), collects sound, amplifying and directing it into the AUDITORY CANAL (B), a pipe ended by the thin skin of the TYMPANIC MEMBRANE (C), or EARDRUM. Vibrations are carried across the air-filled TYMPANIC CAVITY (D) by three tiny bones (the OSSICLES): the HAMMER (E), ANVIL (F) and STIRRUP (G). The EUSTACHIAN TUBE (H) connects the middle ear to the throat, acting as a valve to equalise pressure differences. The stirrup attaches to the liquid-filled inner ear via the FENESTRA OVALIS (J), or 'oval window'. A thin membrane below, the FENESTRA COCHLEAE (K), or 'round window', enables the liquid to move, allowing vibrations to enter the spiral COCHLEA (L, Latin for snail). Conjoined are three SEMICIRCULAR CANALS (M) that monitor balance and motion. Within the cochlea lies the mysterious ORGAN OF CORTI (N), with its elaborate arrangement of hair cells, which sway in the vibrating liquid, triggering nerve impulses to the brain.

The human ear can detect frequencies ranging from between 20 Hz and 20,000Hz, however other animals are capable of hearing well beyond that (*see opposite*).

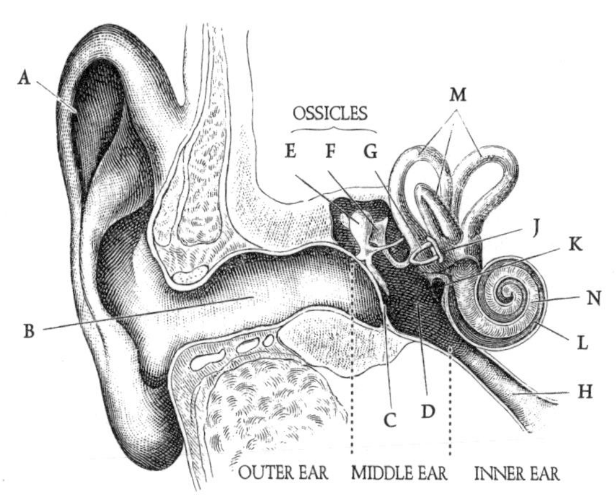

PARTS OF THE EAR

A - PINNA
B - AUDITORY CANAL
C - TYMPANIC MEMBRANE
D - TYMPANIC CAVITY
E - HAMMER
F - ANVIL
G - STIRRUP
H - EUSTACHIAN TUBE
J - FENESTRA OVALIS
K - FENESTRA COCHLEAE
L - COCHLEA
M - SEMICIRCULAR CANALS
N - ORGAN OF CORTI

ABOVE: *The marvels of the human ear. The complex shape of the pinna encodes directional information, whilst the ossicles mechanically match the differing impedances of air and liquid. Hair cells within the Organ of Corti respond to high frequencies at the wide opening of the cochlea, with low frequencies being picked up at the narrow end.*

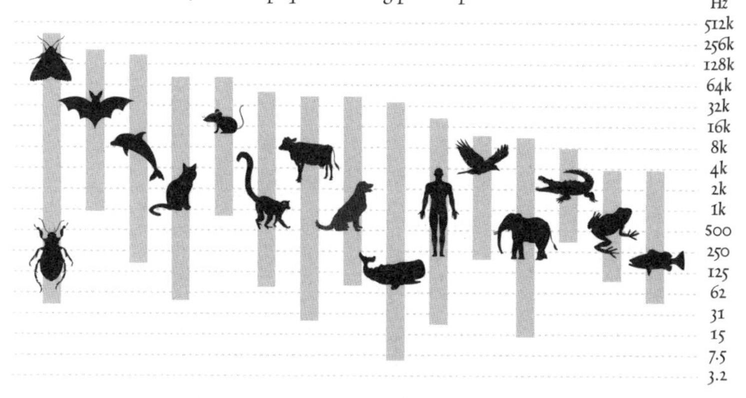

ABOVE: *Comparative frequency ranges. A bat's hearing can extend up to 200kHz, whilst some moths are able to hear as high as 300kHz to avoid them. Elephants have a remarkable range, from 14Hz to 12kHz, and can pick up rumbles from as far as 10 miles away. In the oceans, low frequencies travel even further; the songs of the humpback whale can be heard 100 miles away, and blue whales calling at around 10Hz can hear one another over distances up to 1000 miles.*

LOUDNESS & DECIBELS

from rustles to rockets

The power of sound waves is measured as the average sound energy flowing through a point, in watts per square metre (W/m^2).

The human ear is extremely sensitive and can distinguish sounds differing in intensity by the order of several millions. Sound power density is generally expressed as a ratio of how much louder it is than the quietest sound we can hear, a reference level of 10^{-12} W/m^2. The intensity of sound above this is then measured in DECIBELS (dB), using a logarithmic scale to manage the huge range of intensities. With the quietest audible sound set at 0 dB, a sound 10 times more powerful is 10 dB, 100 times more powerful is 20 dB, and a power ratio of 1,000,000:1 is written as 60dB. Quieter than audible volumes use negative quantities, e.g. a power ratio of 1:100 is -20dB. An orchestra can produce levels that vary from 40 dB right up to 100dB—a million times as much!

The intensity of a sound decreases inversely proportional to the square of the distance from the source. So if you halve your distance to a bird, its song will be four times louder. This translates as an increase or decrease of 6 dB as you halve or double your distance (*below left*).

For telecom equipment, the need for a standard measure of loudness led to the *volume unit* (VU). Derived from the root mean square of a signal, VU meters show an average level over time (*below right*).

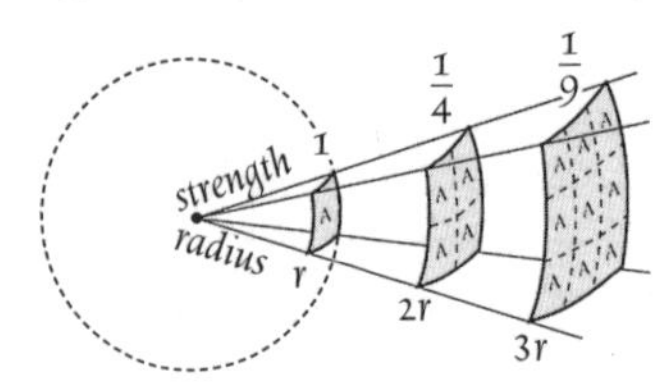

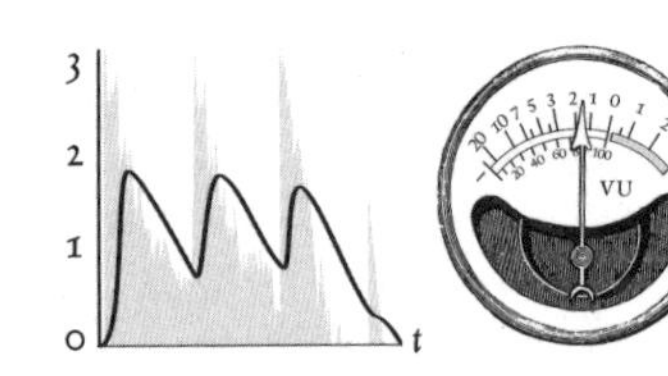

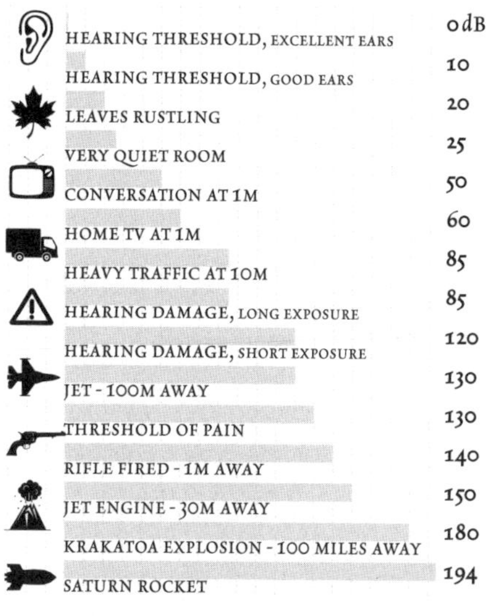

LEFT: The decibel scale, with a range of sample volumes. BELOW: The loudest sound ever recorded was probably the 1883 volcanic explosion of Krakatoa, Indonesia - so loud it was heard 3,000 miles away. The captain of a ship 40 miles away reported that eardrums of half of his sailors burst, and on Mauritius it sounded, "like the distant roar of heavy guns". A vast pressure wave radiated from Krakatoa in all directions. Travelling at the speed of sound, it was detected by barometers in fifty weather stations around the world; three or four circuits of the wave sweeping the earth were tracked over five days, one every 34 hours.

TUNING

an ancient problem

The earliest story of musical mathematics dates back to c.2000 BC. The Chinese Emperor asked court musician Ling Lun to tune a set of bells, so he cut a bamboo pipe to an auspicious length (the FUNDAMENTAL pitch), and a second at two-thirds this length (to give a note 3:2 higher, the FIFTH). A third pipe was cut, equal to the second one plus one third (to give a note 4:3 lower, the FOURTH). By repeatedly applying this 'rule of thirds', going up a fifth and down a fourth, Ling Lun created a series of pipes tuned to twelve equally-spaced notes over two octaves.

Centuries later, the Greek philosopher Pythagoras [c.570–c.490 BC] heard some blacksmiths' hammers ringing in a pleasing way (*opposite right*). Recognising the musical intervals of the octave (2:1), the fifth (3:2) and the fourth (4:3) of the seven-note diatonic scale, he discovered that the harmonious tones were due to the hammers' comparative weights, not their shape or the force applied.

Next, Pythagoras tried stretching equal strings with various weights (*right*). The first string was stretched by 6 units of weight. Another, stretched with 12 units, sounded a note exactly one octave higher. 8 units produced the interval of a fifth; 9 units the fourth. Repeating the experiment using a monochord (*opposite top*), he realised that simple ratios, between lengths, weights, or pitches, are the basis of harmony, and are most pleasing to the human ear.

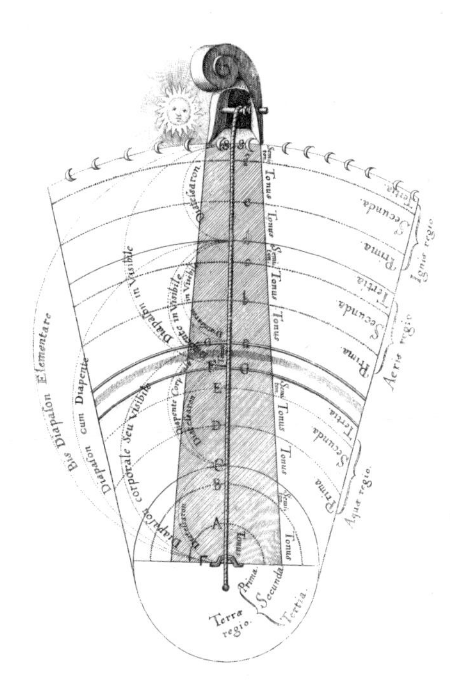

LEFT: *The monochord: a single string of fixed length and tension, tuned by a moveable bridge. When the string is divided into 2, an 'octave' (2:1) is heard. Dividing it into 3 produces the 'fifth' (3:2 or 3:1, the fifth note in the 8-tone scale). 4:3 is the 'fourth' (the fourth note in the scale). The interval between the fifth and the fourth is 9:8 (the 'tone'). A major third is 5:4 and 6:5 is a minor third.*

ABOVE RIGHT: *Medieval depiction of the story of Pythagoras passing the blacksmiths.* BELOW: *A 'Bianzhong' set of 65 bronze bells in a lacquer frame, excavated from a Chinese tomb dated to c.433* BC, *is the oldest known chromatically-tuned instrument in the world (a chromatic scale is a 12-note scale). The bells are tuned to* A, A#, B, C, C#, D, D#, E, F, F#, G, G#, A.

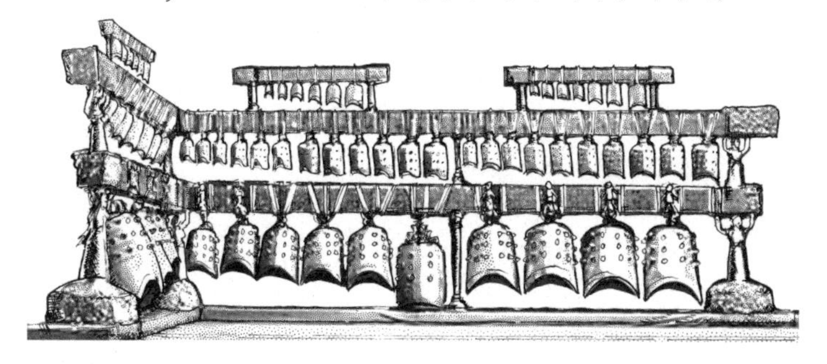

PERIODICITY & PITCH

from low to high

A note vibrating twice the frequency of another sounds 'the same but higher'. In between these can be placed other notes to form a SCALE. Greek scales tended to use seven unequally-spaced notes, with the one completing the cycle known as the OCTAVE. However, the 'octave' also completes the five-note pentatonic scale and the twelve-note chromatic scale. Using only the fifth (3:2), Pythagoras constructed a CIRCLE OF FIFTHS (*below*), a chromatic scale similar to Ling Lun's.

From a starting note, divide a string into three. A string length two-thirds of the original sounds a frequency 3:2 of the original, up a fifth (an octave below the third harmonic, 3:1). Starting at C, and moving up in fifths gives G, then D, then A, then E, B, F#, C#, G#, D#, A#, F and C. The sequence appears to finish seven octaves higher than the start. However, the pitch of the final note slightly overshoots seven octaves: a difference known as the PYTHAGOREAN COMMA (*below*).

The mismatch between octaves and fifths meant that instruments needed to be tuned to (and played in) a particular key. The problem was unresolved until the 18th century with the adoption of EQUAL TEMPERAMENT, widely used today, where each octave is divided into twelve equal intervals. Apart from the octave, all perfect harmonies vanish, but the system is a workable compromise (*see page 19*).

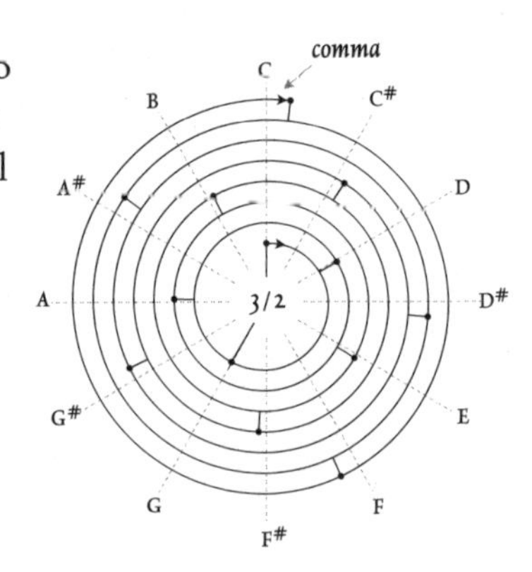

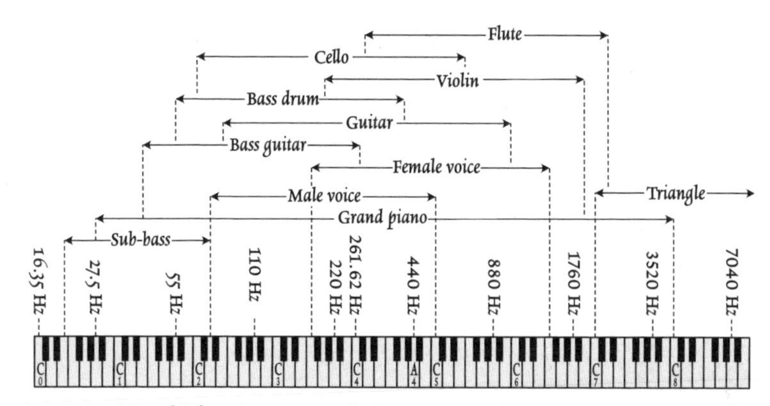

ABOVE: *The pitches of different instruments and voices. International Standard concert pitch sets* A_4 *(above middle* C_4*) at* 440Hz. *However it can vary from a Baroque* 415Hz, *to the special properties of* 432Hz, *up to the* 18*th* C. *Chorton pitch of* 466Hz. *The relationship between pitch and tension (next page) was formalised in equations by* Marin Mersenne [1588 – 1648], *after work by* Galileo Galilei [1564-1642].

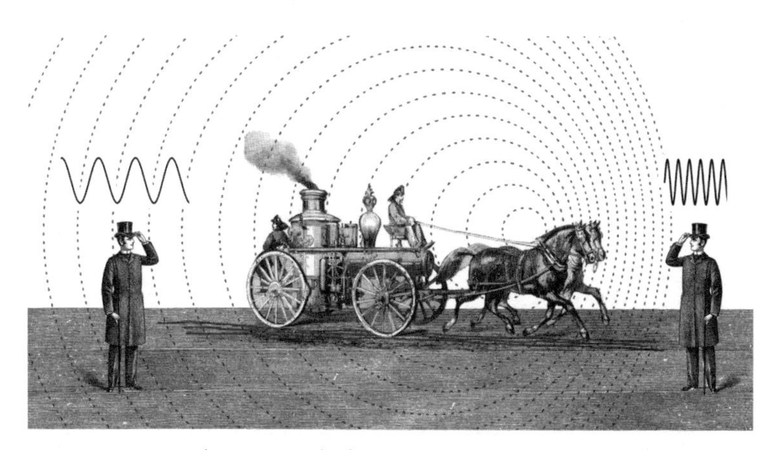

ABOVE: Doppler *shift, an acoustic effect first described by physicist* Christian Doppler *in* 1842. *Here, a moving sonic source (a police siren, fire engine bell or hoofbeats) approaches an observer, causing successive wave crests to bunch up into a higher experienced frequency, until the source passes, after which the waves become stretched out into a lower pitch as the source moves away.*

STANDING WAVES IN STRINGS

nodes & antinodes

When a tensioned string, anchored at both ends, is plucked, two wave pulses move out from the source of the vibration and reflect back from the ends. Very shortly, there will be waves travelling simultaneously in both directions along the string—a standing wave. The ends of the string form NODES, still points with no vibration. The longest standing wave, the length of the string, is the first mode of vibration, and generates the fundamental frequency.

A guitar string of fundamental pitch 1, lightly touched halfway, and plucked elsewhere, will produce the second harmonic, or octave above (2:1). Touching the string at a third of its length, and plucking elsewhere, produces the third harmonic (3:1), and so on.

Other smaller standing waves exist within the string, as long as there are stationary nodes, A and B, at each end. These have wavelengths that are integer-related to the length of the string: $\frac{1}{2}$, $\frac{1}{3}$, $\frac{1}{4}$, $\frac{1}{5}$ etc, and produce the HARMONICS of the fundamental (*below and opposite*).

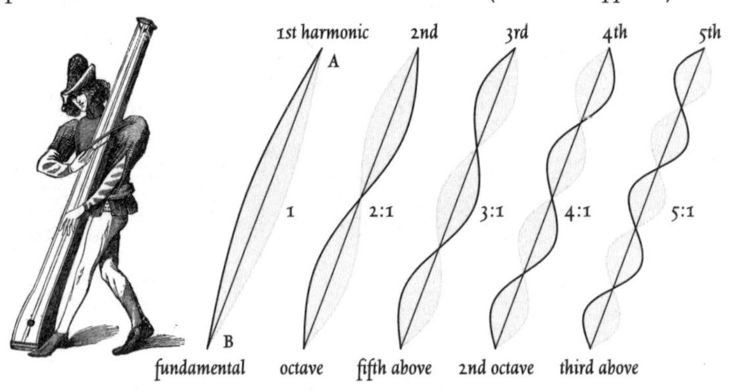

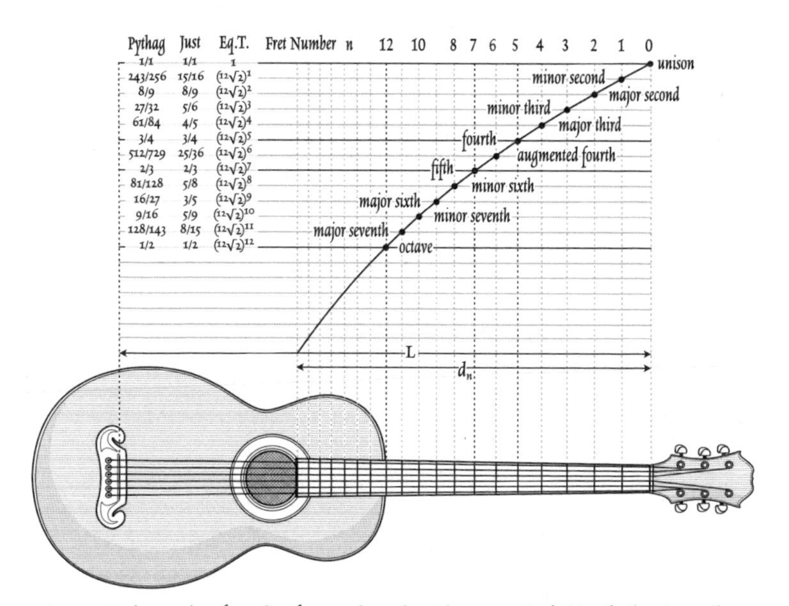

ABOVE: *Various tunings for guitar frets. In the early 16th century, the fashion for hearing perfect thirds (5:4, the major third, and 6:5, the minor third), led to Just Intonation (and slightly different fret positions). The widespread adoption of Equal Temperament from the late 18th century, created a new tuning system where each note varies from its neighbours by* $\sqrt[12]{2} = 2^{1/12} = 1.059463\ldots$

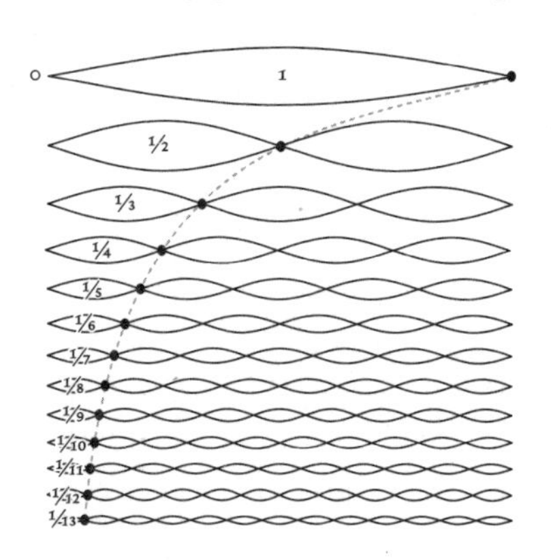

Equal Temperament permits the "Luthier's Rule of 18" (more accurately, C = 17.817). For a string length (L), the position (d_n) of a fret (n) can be calculated from the previous fret:

$$d_n = d_{n-1} + \frac{L - d_{n-1}}{C}$$

The frequency (f) of a string depends on its length (L), tension (T), and mass per unit length (M) according to Mersenne's Formula:

$$f = \frac{\sqrt{T/M}}{2L}$$

LEFT & FACING: *The harmonic series.*

Standing Waves in Pipes

open & closed

The air inside a long, thin cylinder open at both ends can oscillate relative to the air outside, but not at the two open ends, which become nodes. A standing wave forms inside the pipe, with the wavelength of its fundamental frequency equal to the length. Just as with a vibrating string, other standing waves coexist inside the pipe; with nodes at both ends, they will be integer-related. The flute is effectively a pipe open at both ends (*opposite, left-hand images*).

Sealing one end means air can only oscillate at the closed end, creating an ANTINODE. As the open end is a node, only half a standing wave forms, with a pitch half the frequency of the open pipe, an octave lower. Consequently any other standing waves must also fit between the node and the antinode, allowing only odd-numbered multiples of the fundamental: $\frac{1}{3}$, $\frac{1}{5}$, $\frac{1}{7}$, etc. The clarinet works in this way, like a pipe with one closed end, the odd-numbered harmonics giving its distinctive hollow-sounding voice (*opposite, right-hand images*).

Brass instruments rely on harmonics for producing their notes. The bugle, for instance, has no valves to change the length of tube and hence the pitch: tunes are played solely by picking out the different harmonics by blowing harder and softer (*below*).

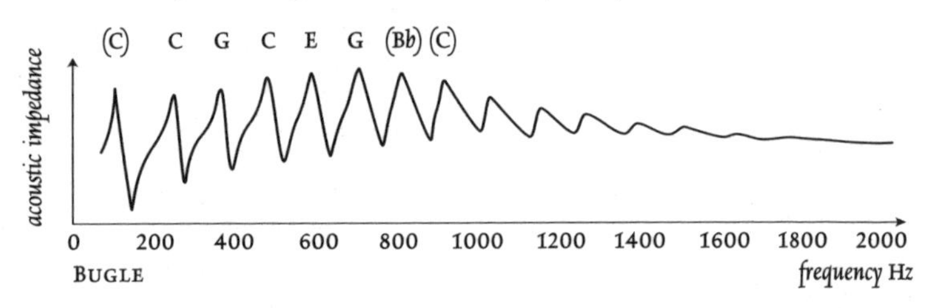

Open Pipes

L

node

antinode

fundamental / 1st harmonic wavelength $\lambda = 2L$

2nd harmonic / 1st overtone $\lambda = \frac{2L}{2}$

3rd harmonic / 2nd overtone $\lambda = \frac{2L}{3}$

4th harmonic / 3rd overtone $\lambda = \frac{2L}{4}$

Pipes open at both ends, e.g. flutes, produce both odd and even harmonics.

Closed Pipes

L

motion

pressure

fundamental / 1st harmonic $\lambda = 4L$

3rd harmonic / 1st overtone $\lambda = \frac{4L}{3}$

5th harmonic / 2nd overtone $\lambda = \frac{4L}{5}$

7th harmonic / 3rd overtone $\lambda = \frac{4L}{7}$

Pipes closed at one end produce only odd harmonics and sound an octave lower.

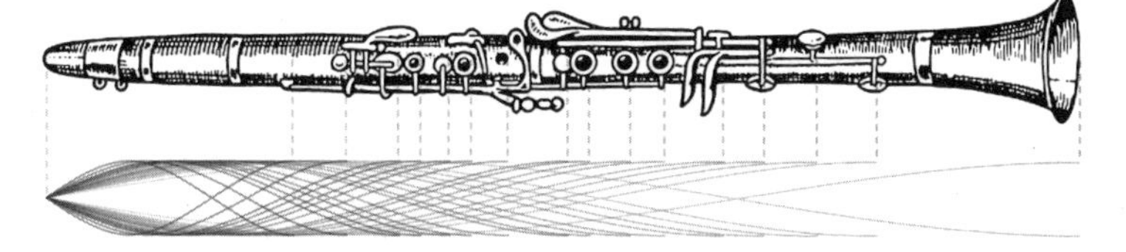

Above: *The fundamental frequency of a pipe is governed by the length of the vibrating air column. For woodwind instruments, such as the clarinet above, the effective length is changed by holes in the pipe. In reality, the position, size, depth, air pressure and number of holes covered all contribute to final pitch.*

HELMHOLTZ & RESONANCE

ear today, gong tomorrow

Any acoustic system will RESONATE, amplifying soundwaves which match its own natural frequencies. Singing into a piano whilst holding the sustain pedal gives good example: several strings will vibrate in sympathy, not just at the sung frequency, but also those tuned to the same note in other octaves, above and below. Similarly, a twanged tuning fork will cause an identical fork nearby to vibrate (*below*).

Many insights into the nature of sound were made by Hermann von Helmholtz [1821–1894]. To analyse a sound, he fitted a succession of tuned HELMHOLTZ RESONATORS in his ear to estimate the strength of the various harmonics (*opposite top*). Using this method he discovered the non-harmonic component frequencies of bells and gongs.

The interplay of internal resonances determines the sound produced by an object. Musical instruments are carefully shaped to accentuate some vibrations whilst dampening others. Resonators can be also be used to treat rooms by absorbing unwanted frequencies (*lower opposite*).

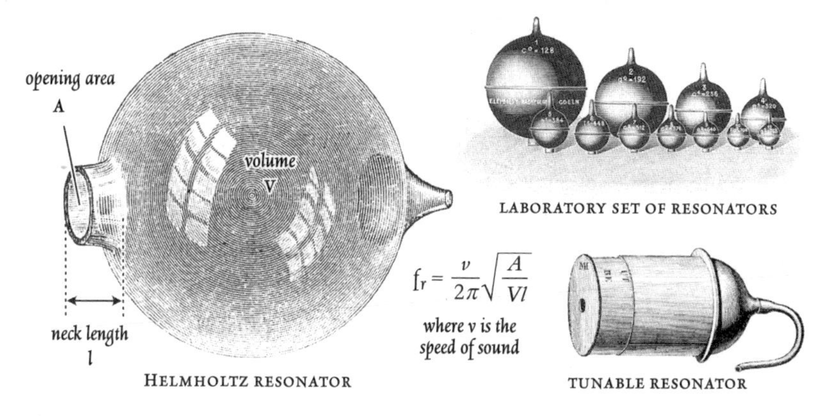

ABOVE: *Helmholtz resonators are made of glass or metal, and have two openings; one shaped to fit the ear, the other having a short neck where sound enters. Resonators collect a specific sound frequency (f_r) according to their size. They can be used to discover or dampen specific frequencies. Tunable resonators act a little like a trombone in reverse, with a sliding mechanism to change the resonant frequency.*

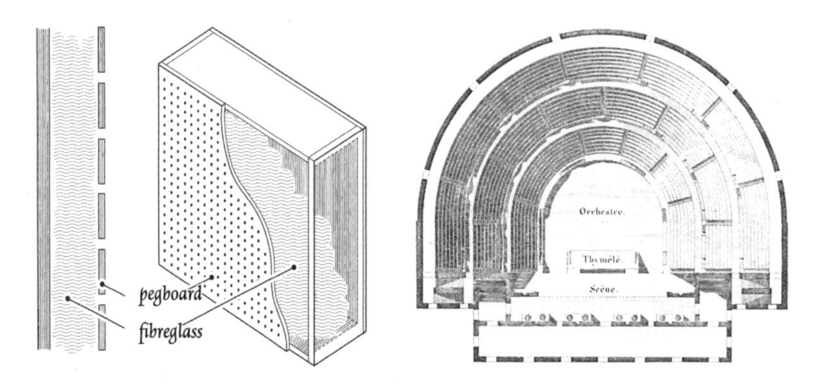

ABOVE LEFT: *An acoustic panel designed to act like multiple Helmholtz resonators. Packed with an absorbent material, it deadens specific frequencies by converting sound energy into heat.*
ABOVE RIGHT: *Clay pots, sometimes filled with ashes, have been found under the seats of several Greek theatres and in the walls of medieval churches. These may have been used to help acoustic response.*

COMPLEX WAVES
from simple things

In 1822, the French mathematician Joseph Fourier [1768–1830] discovered that all waveforms, no matter how complex, can be broken down into a series of basic sine waves of differing amplitude, frequency and phase. This insight led to the FAST FOURIER TRANSFORM, a powerful mathematical technique that dismantles waves to reveal their individual frequency components (*opposite top*).

When waves collide, interesting things happen. Two notes that are very close to each other interfere to produce a pulsing BEAT FREQUENCY (*opposite centre*).

Simple waves can be combined to create richer sounds. The stops of a church organ are an early instance of additive synthesis. In the same way, modern synthesizers can build up sounds from oscillators, digital samples or wavetables to give a huge range of sonic possibilities (*right*). The fluttering of a bird's wings is the result of AMPLITUDE MODULATION (AM), where the intensity of one signal is varied by a second. Altering the frequency of one wave by another causes FREQUENCY MODULATION (FM), like a violinist's vibrato. FM based synthesisers are particularly adept at imitating the shifting harmonics of chimes and bells (*lower opposite*).

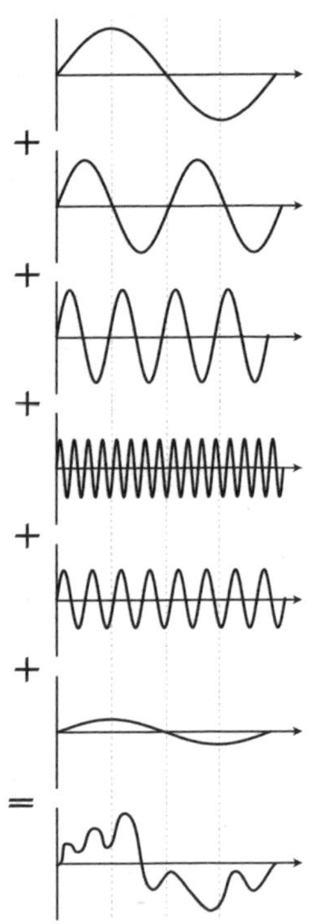

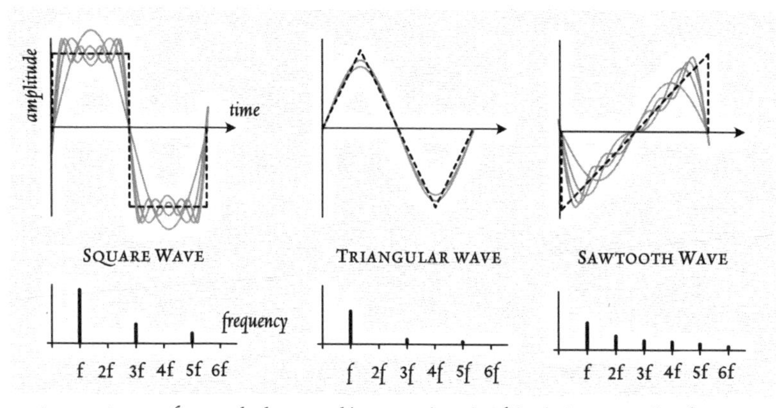

ABOVE: *Any waveform can be decomposed into a Fourier series of simple sine waves. These frequency graphs show the relative strengths of the fundamental (f) and successive overtones after applying fast Fourier transforms. Note that the square and triangular waves have no even harmonics.*

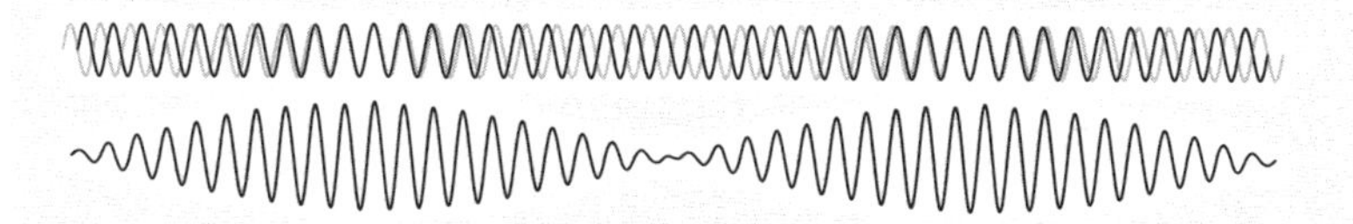

ABOVE: *Beating (for example the warbling of a piano accordian) occurs when waves close to the same frequency cause constructive and destructive interference. The beat frequency is the difference between them - e.g. sounds of 175 Hz and 179 Hz will produce a beat frequency of 4 Hz.*

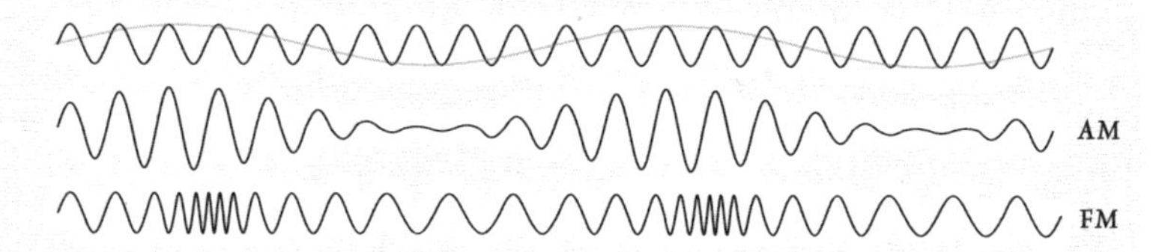

ABOVE: *In amplitude modulation (AM), the size of the first wave (black) is made proportional to the second wave (grey), while its frequency remains constant. In frequency modulation (FM), the frequency of the black wave is varied by the second, while its amplitude stays the same.*

OVERTONES & PARTIALS
timbre & formants

Many simple sounds consist of a fundamental frequency along with a series of harmonics—integer ratio overtones (*see page 18*). More complex sounds may have additional, non-integer related overtones that are termed PARTIALS. Other sounds, such as bells, are rich in INHARMONIC PARTIALS—overtones in no musical relationship.

Musical instruments, animal sounds and human voices owe their unique TIMBRE, or tonal quality, to FORMANTS—narrow bands of resonant frequencies that are boosted or reduced by the physical properties of the source—similar to playing a recording through a graphic equaliser with some of its sliders pushed up (*opposite top*).

Formants crucially contribute to every instrument's characteristic timbre. In the case of the violin, formants make the difference between a cheap fiddle and a priceless Stradivarius. The raw sound of a bowed string is a sawtooth wave, a fairly unpleasant buzzing that is transformed by the violin's many resonances at different frequencies. The formants are defined by multiple factors, such as the instrument's size and shape, the type of wood used, the varnish, and so on. The image (*below*) shows the vibrational response of a violin's soundboard as different notes are played.

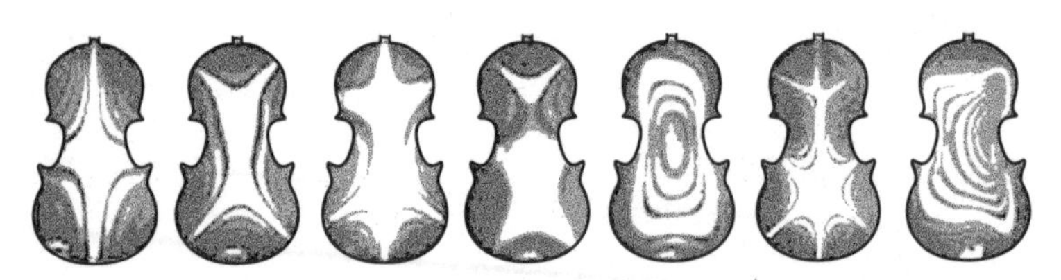

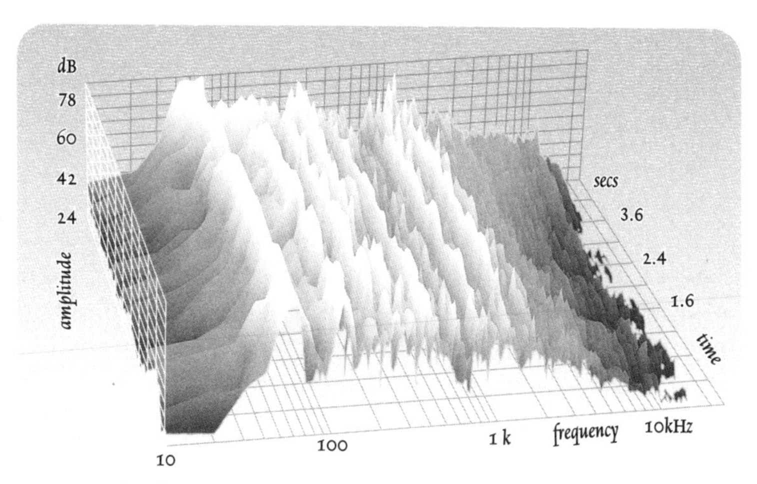

ABOVE: *A time/frequency graph. The rich tonality of acoustic instruments is due to interacting resonances. Subtle shifts occur across the frequency spectrum as a sound evolves through time.*

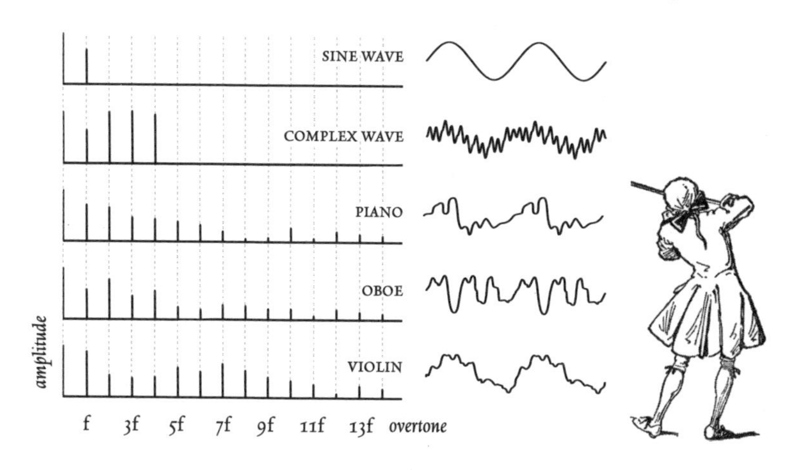

ABOVE: *The comparative strengths of successive overtones of various instruments, in multiples of the fundamental frequency, with two cycles of the associated waveform.*

THE VOICE

in sweet music is such art

The human voice is astonishingly versatile and able to imitate most musical instruments, including synthesizers. Varying the tension of the vocal chords alters pitch, but the final sound is shaped by the mouth, throat and nose. In all, five resonators create the formants of the voice, some of which are continuously variable.

The formant characteristics of these cavities shape the vowels that produce intelligible speech (*lower opposite*). For example, a whisper can be understood perfectly, yet is made without the vocal chords.

Some vocal formants are fixed, whilst others can be altered—the resonance of the mouth in particular can be finely tuned over a wide range. We unconsciously detect the fixed formants, and use them to estimate the speaker's size. If a recorded voice is sped up or slowed down, the formant frequencies also shift, creating the impression that a giant or pixie is speaking, an effect known as 'munchkinisation'.

Overtone, or throat, singing uses the vocal chords to make a low drone note. The singer then manipulates the vocal cavities to pick out individual upper harmonics. A skilled virtuoso can create melodies with the high harmonics, enabling them to effectively sing two notes at once (*or three or four, as shown right*).

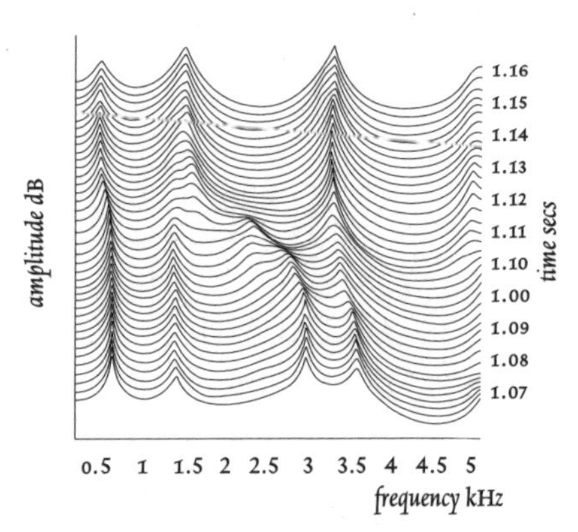

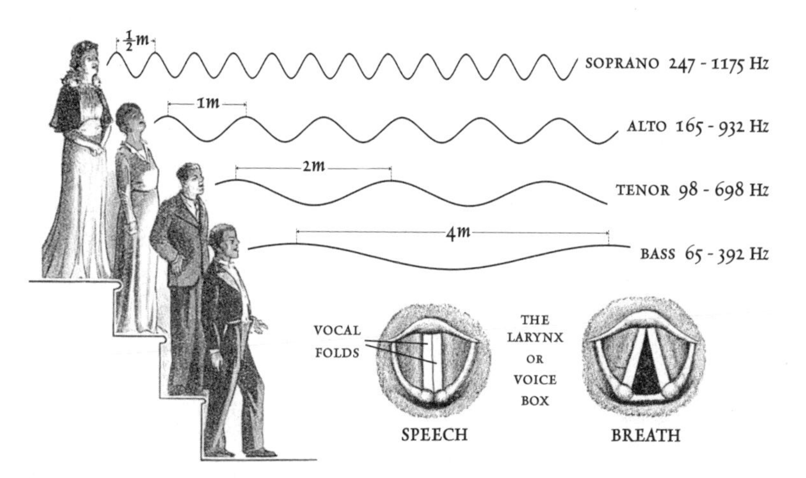

ABOVE: *The fundamental frequency of a female spoken voice is about 200 Hz, with a wavelength of around 1.7m. Male voices, centred about 100Hz, have a wavelength of roughly 3.4m. By contrast, clinking glasses at around 3000Hz, have a wavelength of only a centimetre or so.*

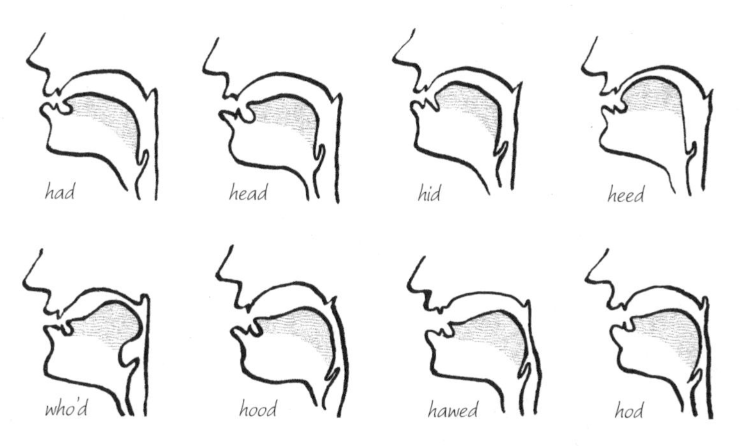

ABOVE: *Vowel sounds arise from the formants produced by altering the shape of mouth, throat and tongue (shaded). Individual voices owe their unique timbre to these interacting resonances.*

Echolocation

location, location, location

Some animals perceive using ECHOLOCATION, emitting sounds and detecting their reflected echoes to sense the world around them.

Bats can see as well as we can, but use echolocation to see at night (*opposite top*). Toothed whales, including dolphins, emit high frequency clicks to hunt and navigate, focusing sounds through the MELON, a fatty organ in their foreheads (*below*). Dolphins can detect prey a 100m away using pulse trains of up to 600 clicks per second. A few nocturnal birds and shrews also send out calls in the dark.

The sonic world of the ocean is a complex interplay of temperature, pressure and salinity, as these affect density. Beneath the warm surface lies the THERMOCLINE, a zone where the sea cools rapidly with depth, causing sound to slow down. At a depth of about 1km the temperature stabilises. Deeper still, the pressure rises and sound speeds up again. This speed inversion creates two acoustic waveguides, the SURFACE DUCT and the DEEP SOUND CHANNEL, which allow sounds like the eerie low swoops of whalesong to carry for thousands of miles (*lower opposite*).

Humans use echolocation devices to detect submarines, shoals of fish, wrecks, currents and mid-oceanic mountain ranges. Electronic SOUND NAVIGATION AND RANGING (SONAR) sets send pings of between 10kHz to 50kHz to bounce back off underwater objects.

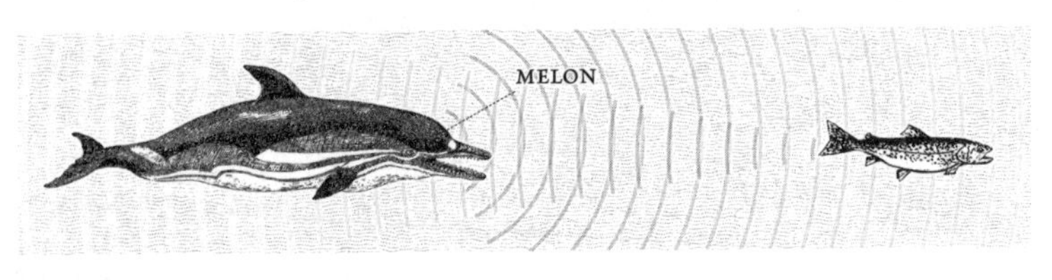

ABOVE: A bat hunts using echolocation, emitting clicks of up to 110kHz which reflect off nearby objects. Echoes of the clicks return and are processed by the bat's brain. Specific neurons recognise the differing harmonics caused by Doppler shifts between call and echo, enabling bats to know the direction, distance, trajectory and physical characteristics of their prey and surroundings.

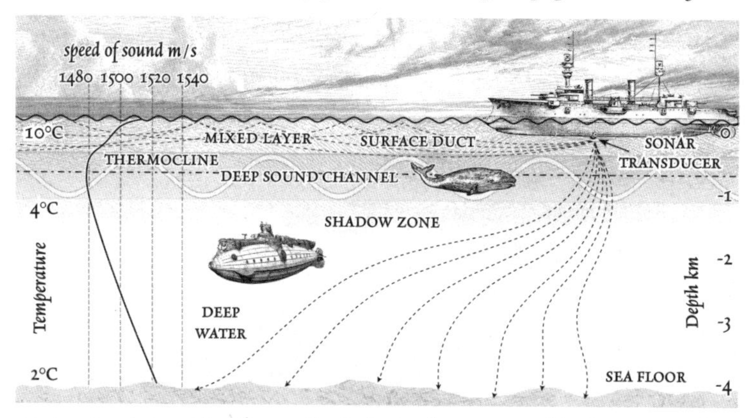

ABOVE: Sound at sea. Many factors affect the density of sea-water, resulting in variations of the speed of sound. The speed inversion caused by the thermocline creates the acoustic wave guides of the surface duct and deep sound channel. The deeper it gets, the faster sound travels, resulting in non-linear wave paths (dotted) and a sonar shadow zone-ideal for hiding submarines.

REVERBERATION
quiet reflections

Just as light reflects in a mirror, sound reflects off hard flat surfaces, bouncing in all three dimensions. Every enclosed space has its inherent resonances, the frequencies of which are dictated by its shape and size. This is particularly evident in an empty room with smooth, stiff and parallel walls and floor, such as a bathroom or toilet, or a hall with a flight of stairs, where a single handclap can produce a distinct tone (*below and opposite*). As the sound reflects off the various surfaces, it reaches the ear via different routes, each of differing length. This results in REVERBERATION, or REVERB—a sharp sound extended by many reflections, gradually fading to silence (*see example of a cave, opposite top*).

Reverberation is measured by using a sonic impulse such as a pistol bang, or bursting balloon. The time taken for the reverberation to fall in level by 60 dB is called the *reverberation time* or RT_{60}. A certain amount of reverberating 'live-ness' in a room can have benefits, including enhancing the intelligibility of speech, but too much can create the opposite effect. In practice, the acoustics of most rooms are 'deadened' to an acceptable level by the sound-absorbing qualities of carpets, curtains, furniture and people.

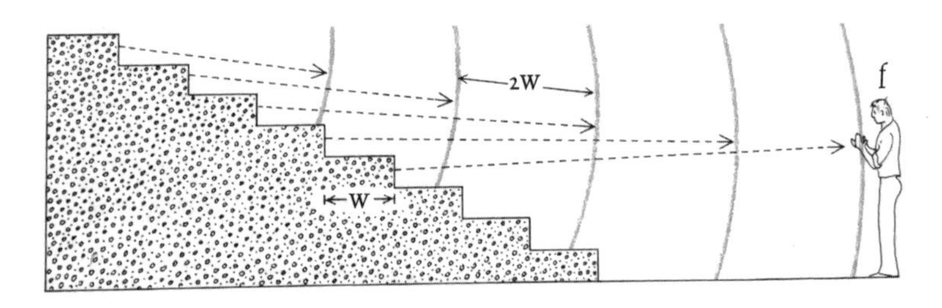

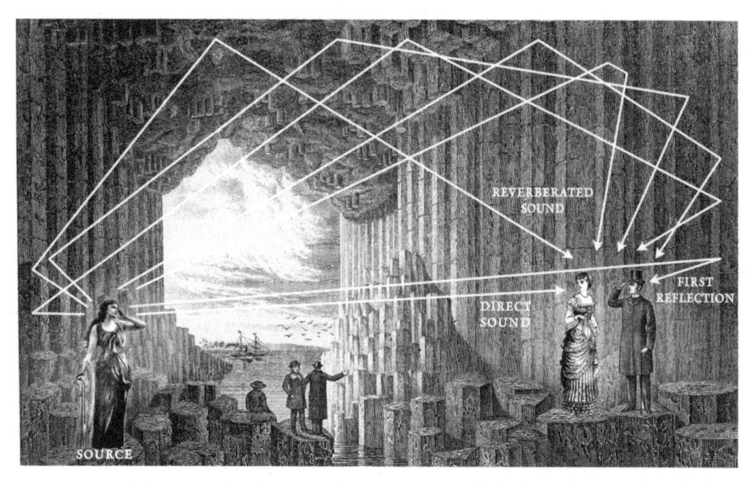

ABOVE: *Reverb in a cave. The direct sound takes the shortest path, followed by the first reflection. After that, multiple reflections are heard with increasing time delay and decreasing intensity. The reverberation time of a room* $RT_{60} = 0.049\ V/a$, *where* RT_{60} = *time for sound to decay by* 60dB; V = *volume of room (in* m^3*); a = total room absorption (in sabins,* *see page 58**).*

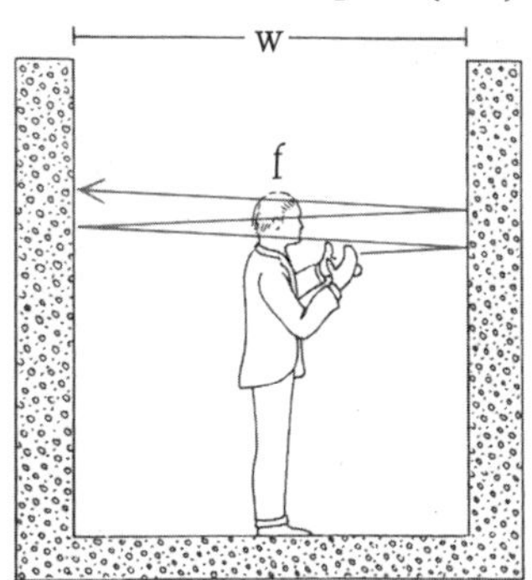

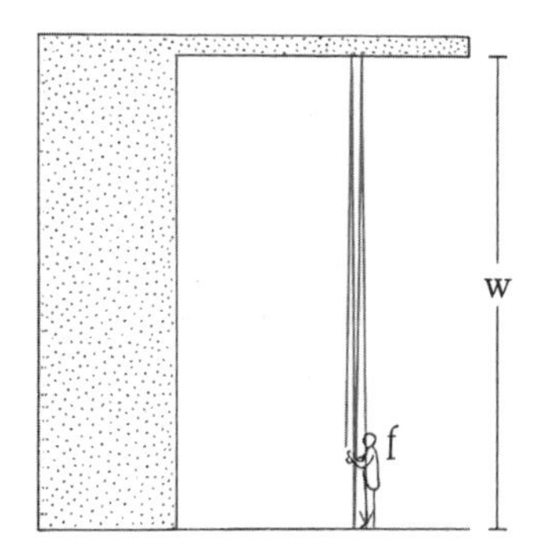

ABOVE LEFT: *Clapping in the centre of a hall width W produces a frequency* $f = C_{20}/W$, *where* C_{20} *is the speed of sound at 20° C, approx 343m/s.* ABOVE RIGHT: *Clapping from the floor under a ceiling produces a tone of frequency* $f = C_{20}/2W$. FACING PAGE: *Clapping in front of a staircase with steps every W metres produces a descending tone, or 'ricochet' of frequency* $f = C_{20}/2W$.

ROOM MODES
unwanted resonances

In an enclosed space, resonances known as **ROOM MODES** are formed as sounds reflect off surfaces to form standing waves in one, two or three dimensions (*below*). Some frequencies are amplified and others are attenuated due to phase-cancellation. Axial modes are the strongest, and may easily be calculated from a room's proportions (*opposite top*). In rooms for recording or listening these resonances can cause problems, particularly at lower frequencies.

Cube-shaped rooms are are especially problematic, having the same resonances in all three dimensions. Rectangular rooms with sides that are multiples of each other, e.g. 3 × 6 m, are almost as bad, since the same series of frequencies will recur. The best sounding rooms have an even distribution of resonant frequencies. Modern recording studios use polygonal plans or length : width : height proportions based on the golden section 1.618 : 1 : 0.618, which dampens all resonant harmonics.

All the objects and surfaces in a room absorb sound at different frequencies (*see page 58*). Low frequency resonances can make a room sound unpleasantly 'boomy' or produce a 'one note bass' effect, where only a single frequency is boosted (*lower opposite*). Large rooms generally sound better than small ones because their lowest resonance tends to be below 20 Hz, and therefore inaudible.

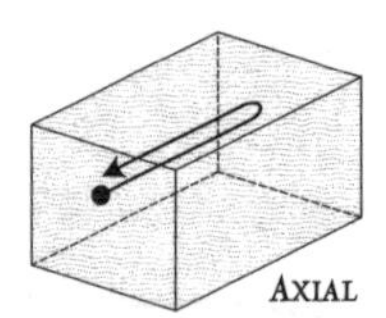
AXIAL

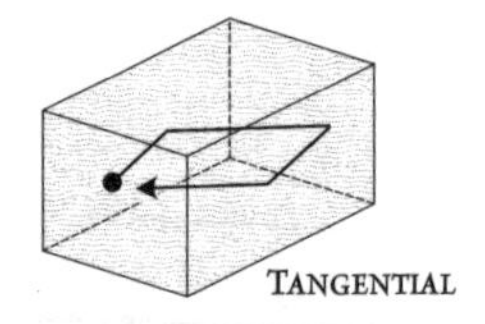
TANGENTIAL

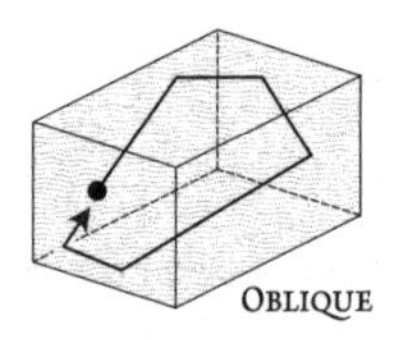
OBLIQUE

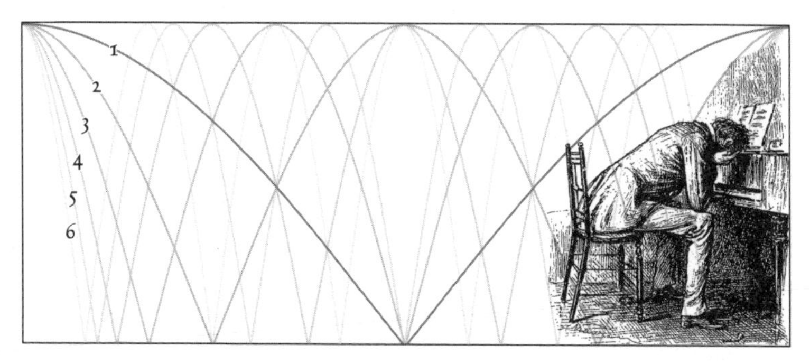

ABOVE: Axial modes. The first resonance is the speed of sound, 343 m/s, divided by twice the distance between surfaces (the sound travels across the room and then back, hence distance is doubled). In a room 4 metres long, the first axial frequency will be 343/8 = 42.8 Hz. The second mode is twice this = 85.7 Hz, the third three times = 128.4 Hz, and so on to infinity, although practically to ten times the first. The process is repeated for width and height.

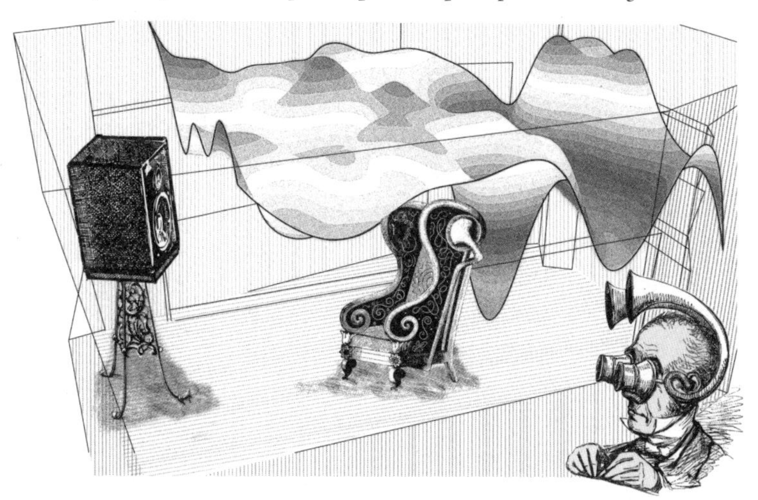

ABOVE: The acoustic character of an irregular room. Test tones are played to map the strengths of different frequencies with a sound pressure meter. Standing waves create nodes and antinodes that seem louder or quieter around the room. The curve represents a single low frequency.

SOUND TREATMENTS

sonic studio spaces

Recording studios are usually 'sound-proofed', both to contain sound within the room and to eliminate noise from outside. Contrary to popular belief, covering the walls with egg boxes does not sound-proof a room, it just reduces the reflection of some high frequencies.

Sound will always escape unless it is dissipated. Mass is the simplest barrier, thus concrete works well, though damping with soft, heavy materials such as neoprene and lead is often more practical.

The best studios have a neutral acoustic, free of resonances and a short reverberation time. Walls are covered with ACOUSTIC PANELS to diffuse sound evenly across the audio spectrum (*opposite top*). Windows between the studio and control room have widely-spaced, non-parallel triple-glazing to eliminate transmission. The very best studios feature a 'room-within-a-room', using a floating floor suspended on neoprene to decouple the studio from outside vibrations.

Corners also tend to focus sound. This can be cured with tall, sound-absorbing cylinders or bunched curtain fabric. Low frequency 'boominess' can be minimised with a BASS TRAP, a large shallow box covered loosely with a soft, heavy sheet such as roofing felt (*right*) and faced with a panel of glass fibre or mineral wool sealed with foil to reflect higher frequencies. Low frequency waves get absorbed and converted to heat.

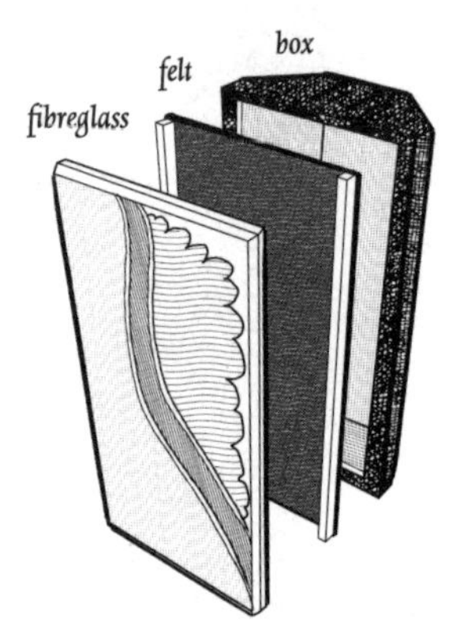

ABOVE: *Acoustic diffusers and panels. Many problems in room acoustics are caused by sound 'ringing' between opposite walls. A few well-placed acoustic tiles can often prevent this.*

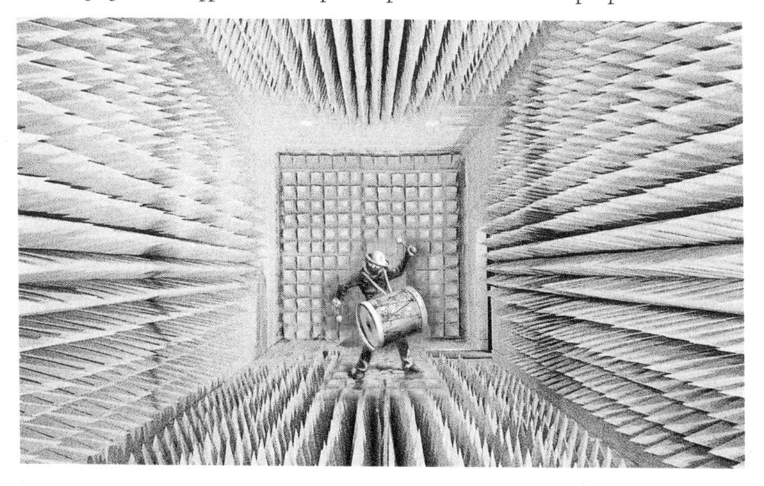

ABOVE: *An anechoic chamber has no sound reflections. Every surface is covered with wedges of fibreglass, as long as possible to absorb low frequencies. Being inside is disturbing; we become aware of our breathing; eventually we notice a hiss, the sound of blood circulating in our ears!*

LOUDSPEAKERS

woofers & tweeters

Loudspeakers convert electrical energy to mechanical energy, and form pressure waves in the air that we hear as sound. There are different types of speakers, but most use a moving VOICE COIL, a coil of wire suspended between the poles of a permanent magnet. The audio signal, an amplified alternating electrical current, passes through the coil, and as the audio signal goes from negative to positive, the coil acts as an electromagnet and moves like a piston as it is attracted and repelled by the magnet. An attached paper or plastic CONE then transmits this movement to the air, producing sound (*below*).

Humans hear sounds from about 20Hz to 20,000Hz, which is too great a frequency range for one loudspeaker to reproduce accurately. For 'high fidelity' a number of speakers of different sizes are mounted in the same enclosure, each reproducing just a portion of the audio spectrum. The audio signal is sent to each speaker via a CROSSOVER, an electronic circuit that restricts its frequency range (*opposite top*).

When wiring multiple speakers, it is crucial to connect them all in phase, e.g. + to red and - to black, or they will cancel each other out.

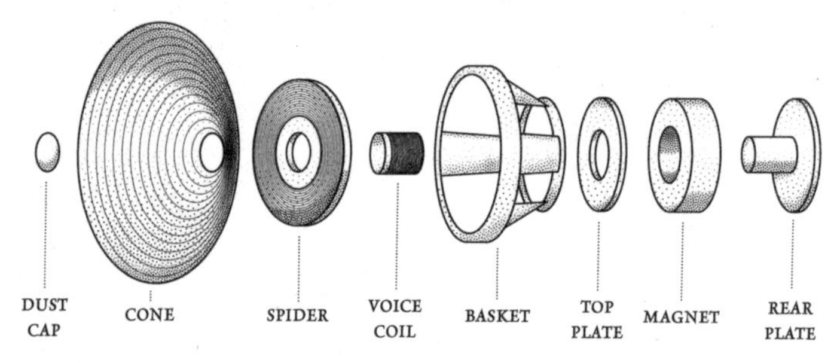

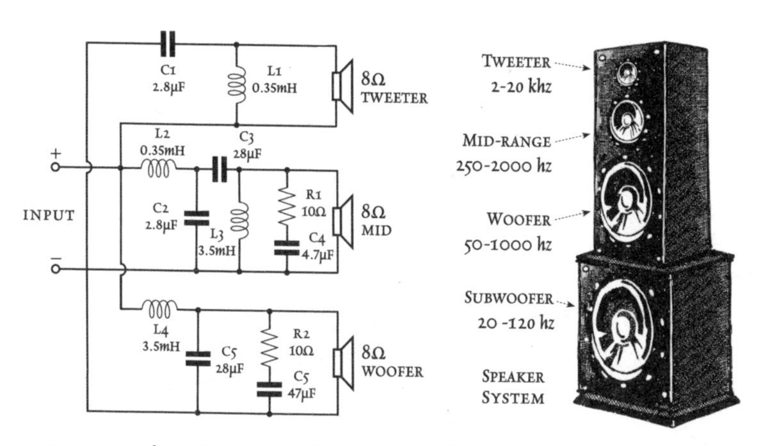

ABOVE: *An electronic crossover. Full-range reproduction needs multiple speakers. The rapid vibrations of high frequencies are handled by small, light tweeters, while bigger woofers with greater mass pump out the bass. The very lowest frequencies are handled by large sub-woofers.*

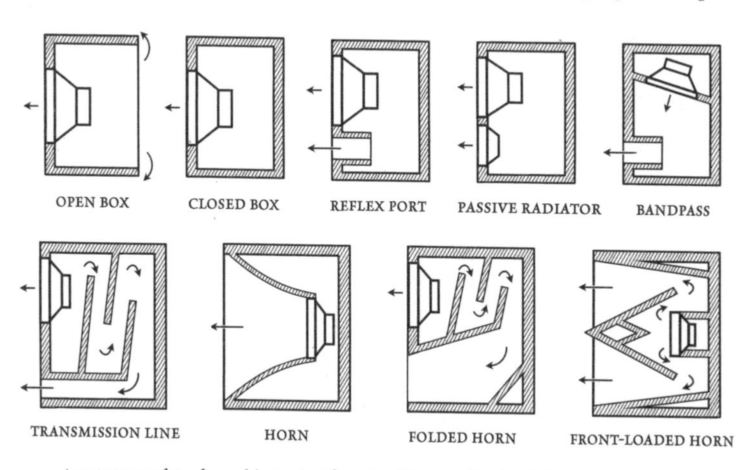

ABOVE: *Loudspeaker cabinets significantly affect sound. Simple boxes can cause phase interference or resonances, so more complex designs use tuned bass reflex ports, labyrinthine transmission lines, and horns to improve efficiency whilst controlling undesirable frequencies.*

PA Systems

addressing the public

How can one person be heard by a crowd? For centuries, the solution was a MEGAPHONE, a conical tube a foot or two long (*see frontispiece*). Megaphones amplify the voice by matching the impedance of the vocal chords to the air, and then directing the sound waves in a particular direction. Larger megaphones are louder—early gramophones used conical horns which could be fifteen feet long in dancehalls.

Sound from a point source radiates in a sphere, its energy dissipating rapidly; doubling the distance from the source results in a quartering of intensity. To help with this, modern PUBLIC ADDRESS (PA) systems stack multiple speakers in a vertical column, so the sound radiates in a cylinder; doubling the distance only halves the intensity. The sound is directed to the audience, not the neighbours. Powerful subwoofers, even placed low on the ground and facing the audience, may still spill some sub-bass, but this can be reduced by adding more subwoofers with reversed phase and aimed in the opposite direction.

For the largest venues and outdoor arenas, extra speakers are often placed out in the audience to supplement the main ones. Since electricity travels at near light-speed, connecting the extra speakers directly to the stage results in the audience at the back hearing the rear speakers first, with the stage sound following as an echo. This can be fixed by slightly delaying the signal so that the rear DELAY TOWER speakers match the direct sound spreading from the stage (*lower opposite*).

SOUND TRAVEL TIME OVER VARIOUS DISTANCES IN DRY AIR AT 20°C:

Distance	1m	5m	10m	50m	100m	500m	1 km	1 foot	5'	10'	50'	100'	500'	1 mile
Time ms	2.91	14.56	29.12	145.6	291.2	1455.9	2915.5	0.89	4.44	8.88	44.4	88.8	443.8	4686.5

ABOVE: A hanging line array of multiple stacked speakers. The J-shape allows the top speakers to focus the sound to reach the back, whilst the lower speakers spread the sound out more at the front. This gives listeners an even intensity of sound no matter where in the audience they are.

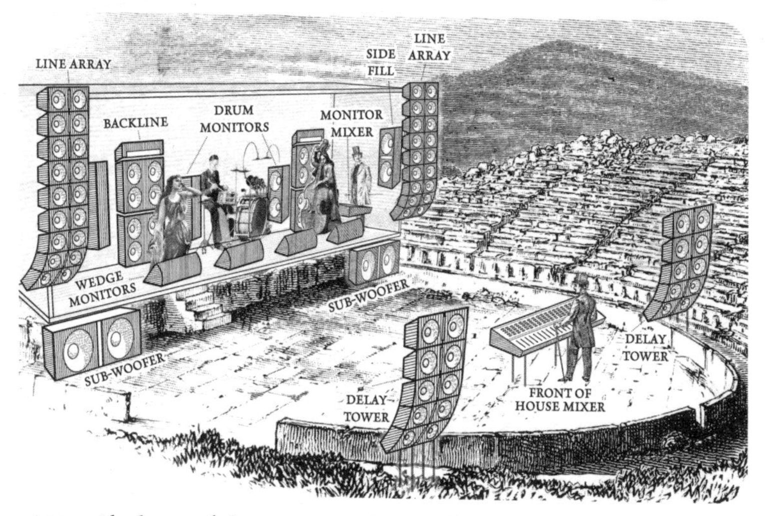

ABOVE: The elements of a large PA system. Line arrays face the audience, with side fill speakers to cover the very front. The sound is mixed by a front of house engineer, away from the stage. To hear on-stage, wedge monitors face the performers, instruments are amplified by the backline, and the drummer is flanked by large monitors, all mixed by a separate engineer to the side.

Hearing in 3-D
where's that noise coming from

To discover where sounds come from, the brain interprets tiny differences between the signals in each ear as spatial information.

Sounds from the side arrive at slightly different levels and times, the nearer ear hearing the signal louder and sooner. In addition, the frequency spectra of sounds are shaped by diffraction, reflection and absorption from the head, torso and pinnae (outer ear folds) before reaching the eardrum. These frequency shifts are expressed by the HEAD-RELATED TRANSFER FUNCTION, (HRTF). By artifically encoding a signal with an HRTF, the brain can be fooled into hearing sounds coming from all directions. The system is not perfect: the brain percieves two sounds less than 40ms apart as one, with both seeming to come from the direction of the first (the HAAS EFFECT). Sounds originating on the cross-section of an imaginary CONE OF CONFUSION are also hard to place, since the time delay between the ears is the same no matter where on the cone they arise (*opposite top*). Finally, the human ear cannot locate sounds below 100 Hz, so only one subwoofer is needed in immersive surround systems such as 5.1, which has five high/mid channels but only one for bass (*lower opposite*).

Three-dimensional sound can be recorded using two microphones spaced an ear's width apart, often mounted in a dummy head (*right*). BINAURAL recordings can be uncannily realistic when heard on headphones, with sounds even appearing to come from above. For greater authenticity, the microphones can be worn over or inside the ears of the recordist's head, or fitted into a actual skull!

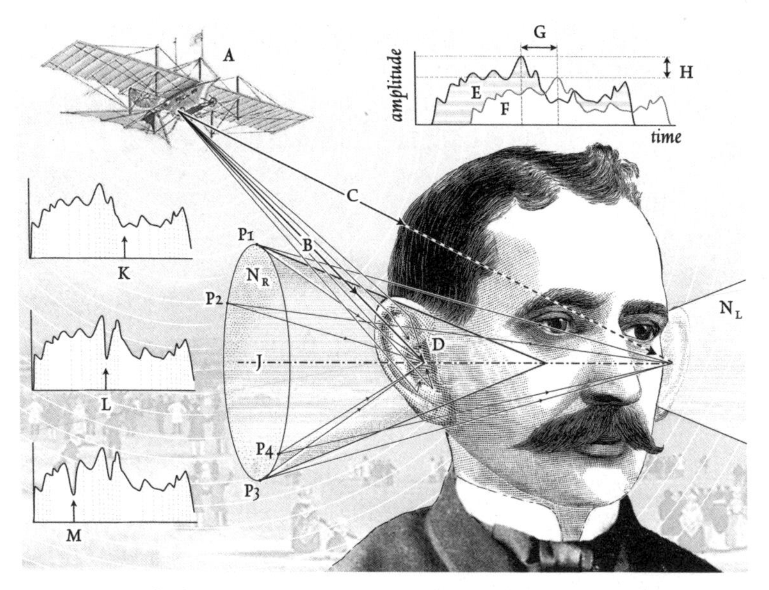

ABOVE: *Hearing in 3-D*. A: *sonic source*, B: *near ear direct sound; shorter distance, higher level;* C: *far ear direct sound, longer distance, lower level;* D: *reflections off pinnae;* E: *near ear signal;* F: *far ear signal;* G: *interaural time difference;* H: *interaural level difference;* J: *interaural axis;* K: HRTF *spectral notch in sounds from above interaural axis;* L: HRTF *spectral notch on axis;* M: *spectral notch below axis;* N: *cones of confusion;* P1-4: *origin points of sounds having equal time differences between the ears.*

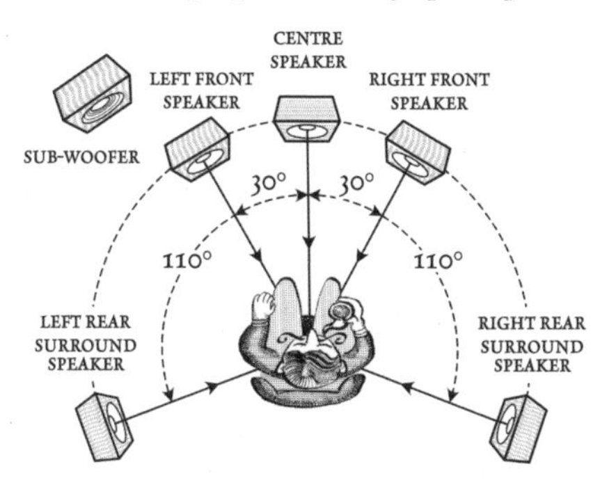

LEFT: *5.1 Surround Sound. Five small satellite speakers are placed around the listener to handle mid to high frequencies, while the bass for all channels is sent to one subwoofer (as the ear cannot place low frequencies, the bass sound seems to come from the satellite speakers). All 'point one' systems work on the same principle, from 2.1 stereo to 12.1 surround and beyond.*

MICROPHONES

testing, testing, one two

A MICROPHONE (or MIC) converts sound energy into an electrical signal.

DYNAMIC, or moving coil, mics are tough and versatile, and are often found on stage. Sound entering the mic vibrates a diaphragm attached to a wire coil suspended between the poles of a permanent magnet, much like a loudspeaker only in reverse (*opposite top*).

RIBBON mics work the same way, but instead of a coil they have a thin, usually corrugated, ribbon of metal foil. Renowned for their high quality sound, ribbon mics tend to be delicate and are easily damaged.

CONDENSER mics can produce very high quality recordings, and are the main type used in studios. They contain a capacitor (also called a condenser) consisting of two thin metal plates that almost touch. One acts as the diaphragm, creating a signal across the plates as it vibrates. Electricity is needed to charge the capacitor, which is provided by the 48 volt phantom power supply from professional studio equipment.

ELECTRET mics are similar to condensers, but use permanently charged battery-powered plates. They can be cheaply made, although some high quality examples have built-in pre-amps and digital recorders.

PIEZOELECTRIC mics exploit the ability of some materials to produce a voltage when under pressure. Mainly used as transducers (tiny beepers for electronics) they can sound surprisingly good as contact mics.

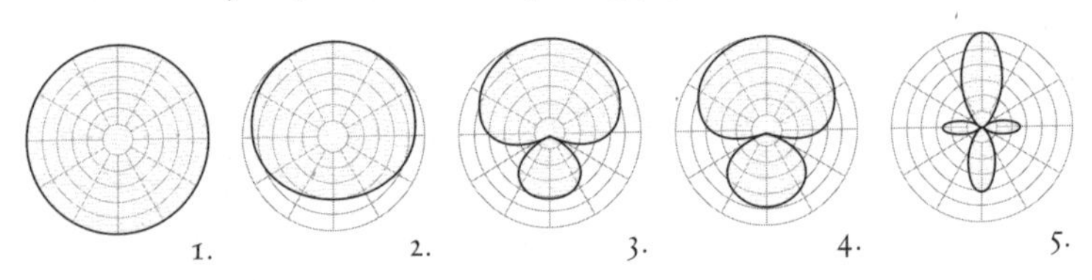

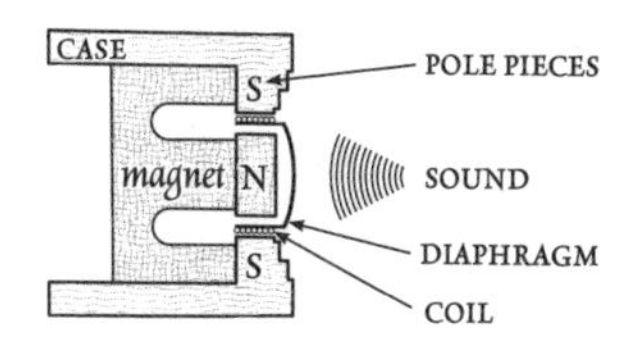

DYNAMIC MICS *often have a heart-shaped cardioid polar pattern (6, below) and exhibit the proximity effect: sounds close to the mic have their low frequencies boosted.*

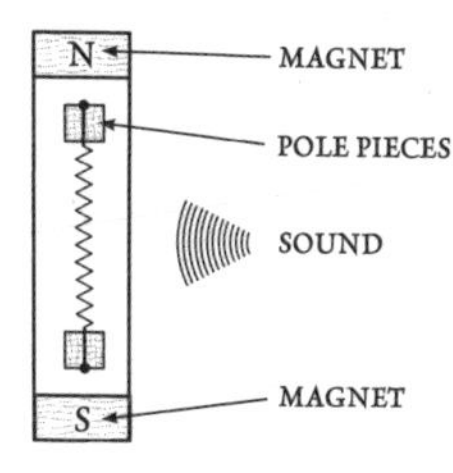

RIBBON MICS *usually have figure-of-eight polar patterns (7, below): sounds recorded at the front and back are of equal level but in opposite phase.*

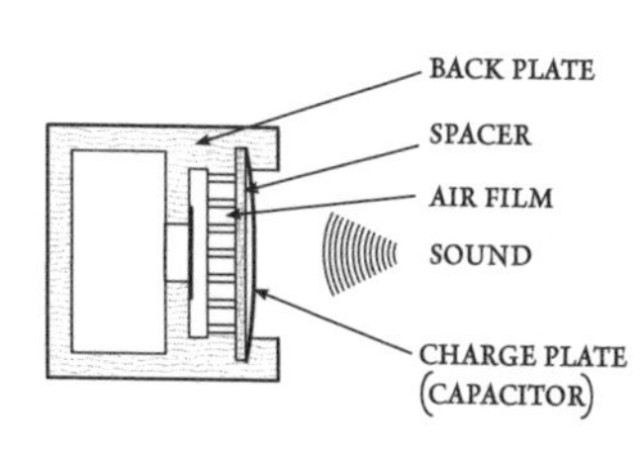

CONDENSER MICS *may feature switchable patterns, often from a spherical omnidirectional response (1, opposite) to hyper-cardioid (4) allowing focus on one direction.*

FACING PAGE & BELOW: *Microphones can be categorized by their three-dimensional directivity, which can affect their frequency response. Common polar pickup patterns are: (1) omnidirectional, (2) subcardioid, (3) supercardioid, (4) hypercardioid, (5) shotgun, (6) cardioid, (7) figure of eight.*

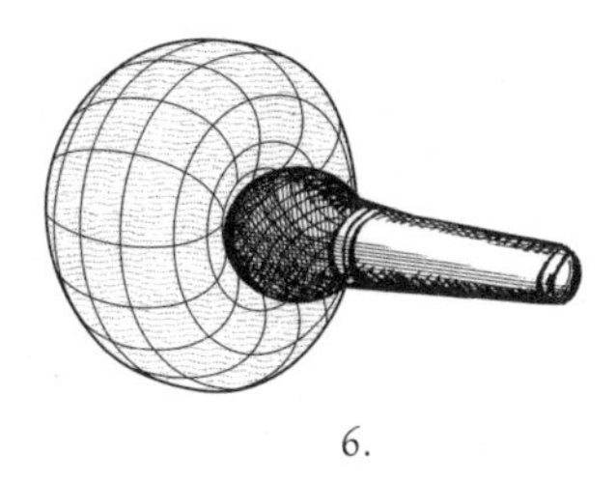

6.

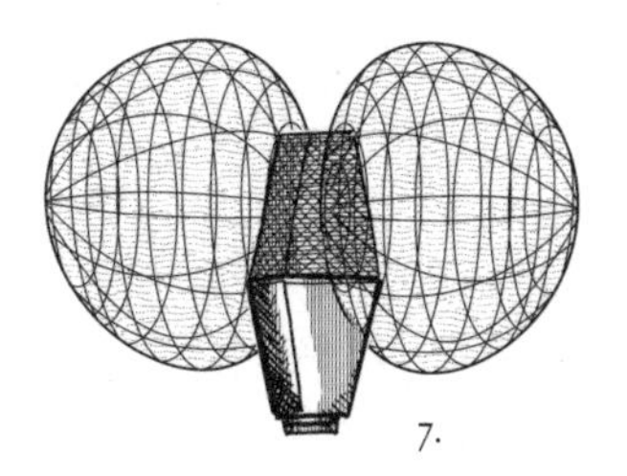

7.

RECORDING
the perfect take

A recording is only as good as the elements that go into it, so getting the best sound from the start is paramount. Early recordings used only one mic. However, from the 1960s, multi-track tape machines allowed multiple simultaneous recordings. As well as enabling more complex arrangements, engineers discovered that putting several mics on each instrument gave a richer, more detailed sound (*opposite*).

For multiple mics, correct placement is essential as phasing issues and odd room tones are hard to correct later. Input gain should be set high enough to avoid hums and hiss, but not so loud that it distorts.

The more information that is captured onto the recording medium, the better the quality. On vinyl and tape this is achieved by running the machine at a faster speed. For digital, this translates to using a higher *sample rate* (the number of bits per second that represent the sound), which needs to be twice the highest audio frequency. Domestic CDs sample at 44.1 kHz, giving a highest frequency of 22 kHz. Studios may sample at 88.2 kHz or more to avoid distorted upper harmonics, a side effect of extensive digital processing.

Wind getting into a mic can cause pops or rumbles. To prevent this, location sound recordists use a 'dead cat'—a fur fabric covered cage that surrounds the mic (*right*). For recording underwater, specialist *hydrophones* are used (or a mic tied up in a condom!).

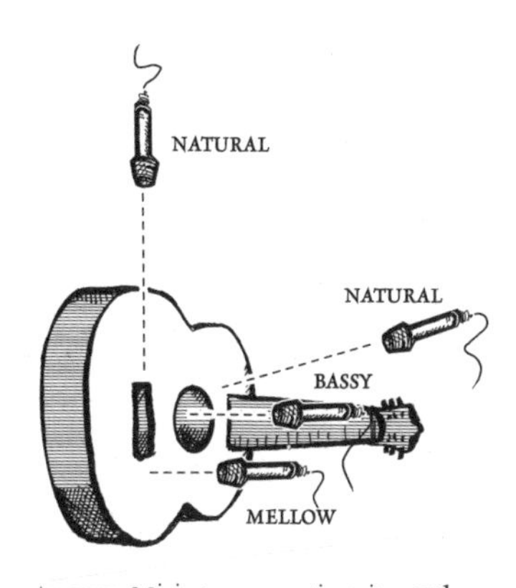

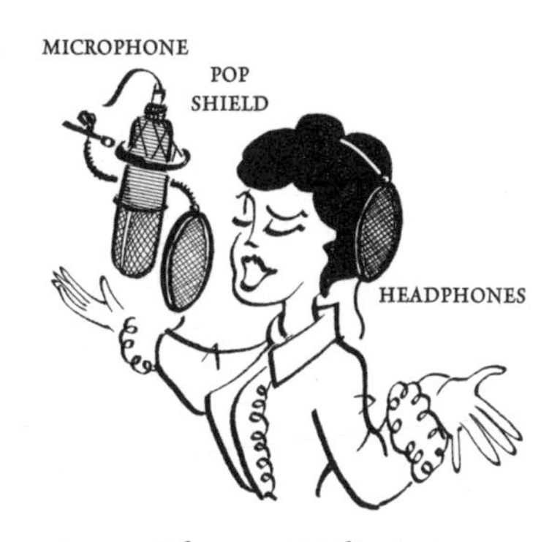

ABOVE: *Micing an acoustic guitar. When recording vocals with guitar, position mics carefully as phase issues can be a problem, when the same wave is recorded with a slight time offset due to a different distance.*

ABOVE: *A fine gauze pop filter protects a sensitive vocal mic from puffs of breath and ensures that the singer is not too close. Closed back headphones allow the singer to hear the music with no overspill to the mic.*

ABOVE: *Mic set up for drums. If there are enough tracks, additional mics on the bottom of the snare and the front of the bass drum will give greater tonal flexibility.*

ABOVE: *Recording a piano. A pair of mics close to the strings give clarity whilst an ambient mic lends richness. Beware of mechanical noise from dampers and pedals.*

EQ & FILTERS

bass, middle, treble

EQUALISATION, or EQ, uses electronic filters to boost or reduce particular frequencies. It can improve odd sounding voices and instruments, remove external noise, and correct poor response from mics and speakers. In dense recordings, the defining sound of each track is often carefully sculpted with EQ to allow each element to be clearly heard (*opposite top*).

A hi-fi's treble and bass knobs are examples of high and low SHELF FILTERS, which work on all frequencies above or below a set point. BAND PASS FILTERS, such as middle controls, work around a central frequency.

The range over which a filter acts can be altered by changing its RESONANCE or Q. Sharp cutoffs are used to take out frequencies completely, such as LOW CUT FILTERS to eliminate traffic rumble, or HIGH CUT filters which block high frequency hiss or digital noise. DE-ESSERS are automatic high cut filters which remove vocal sibilance.

In practice, several EQ elements may be needed to shape a sound; a GRAPHIC EQ divides the spectrum into many discrete bands, covering an octave or less, each with their own control (*below*). A PARAMETRIC EQ allows the user to tune the filter to the problem frequency, and then adjust the Q setting to either cover a broad range, or hone in with a sharp PEAK OR NOTCH on a single area (*lower opposite*).

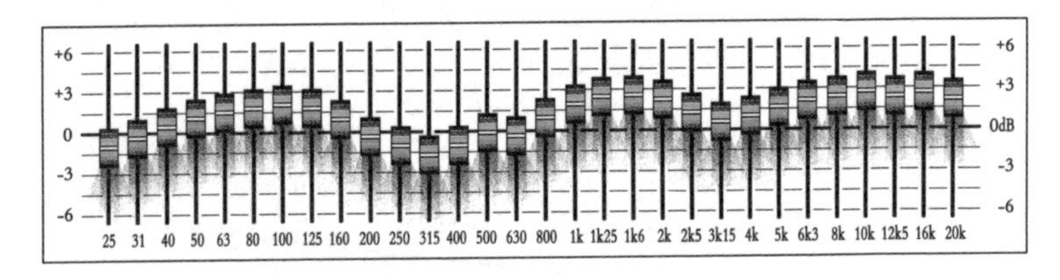

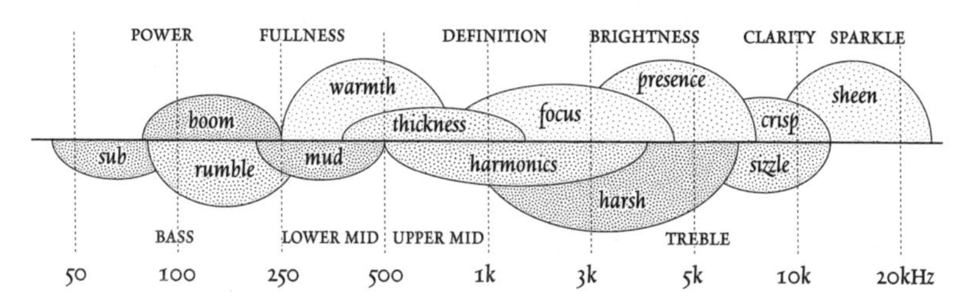

ABOVE: *An EQ cheatsheet—how the ear hears perceives various frequencies.*

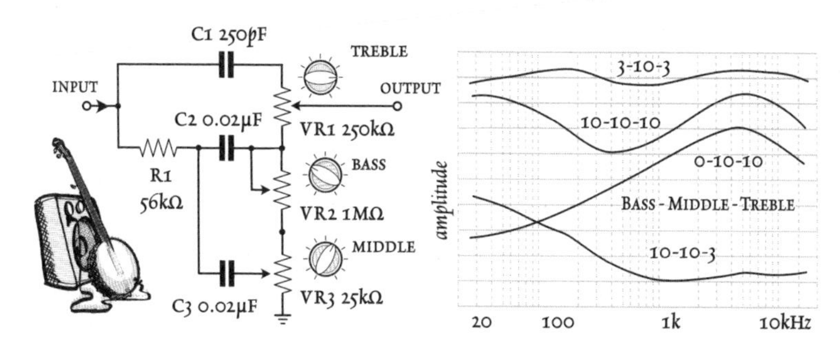

ABOVE LEFT: *a 3-band guitar amplifier EQ circuit.* RIGHT: *the frequency response curves reveal how the highly interactive controls can create a wide range of tonal variation for different settings.*

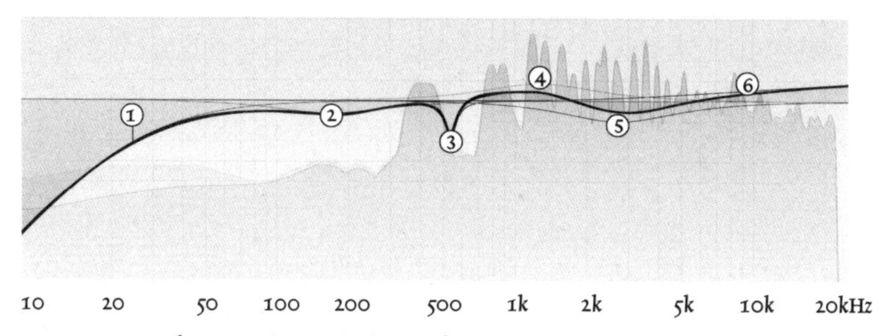

ABOVE: *Many software EQs have multiple bands for precise control: 1. low cut filter; 2. wide Q low-mid scoop; 3. tight Q notch filter; 4. high-mid boost with; 5. compensating dip; 6. high end boost. In specialised linear phase EQs, inter-band interactions are eliminated to prevent frequency distortions.*

COMPRESSION

limiting, gating and expansion

Hi-fi enthusiasts love DYNAMIC RANGE, with the loud crashes really loud, and the quiet pin-drops really quiet. However, for many sound systems the dynamics of an orchestra or rock band are too wide: the quiet parts can be inaudible and the loud parts deafening or distorted.

A COMPRESSOR squeezes the dynamics, raising the level of the quiet parts and reducing the loud ones (*below and opposite*). Used well, compression can be transparent, tightening up vocals and creating impact on drums or bass. When used poorly it can result in audible level pumping and be tiring to listen to. Commercials are often over-compressed, to make every moment of them as loud as possible.

LIMITERS act as very high ratio compressors, often in a *brick-wall* mode to prevent recordings exceeding a maximum level. Here, the limiter is put at the end of a signal chain to instantly reduce any peaks, allowing the rest of the track to be turned up without fear of overloading.

NOISE GATES work as switches, shutting off a signal unless it exceeds a threshold (*opposite mid*). They can stop hums and hisses being recorded. EXPANDERS are more versatile, allowing more and more of the signal through until the threshold is reached. A gated reverb can give a dramatic splash on a drum, then quickly shut it off before it swamps the mix.

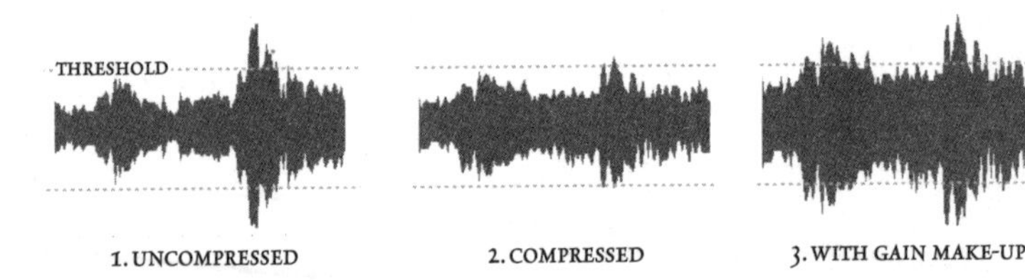

1. UNCOMPRESSED 2. COMPRESSED 3. WITH GAIN MAKE-UP

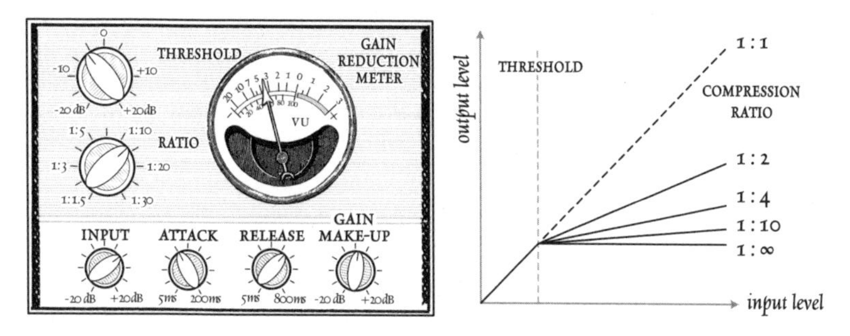

ABOVE: *A compressor. The threshold sets the level at which a signal starts being squeezed. The ratio sets the amount of compression: the higher the ratio, the more gain reduction is applied. As the compressed signal is lower than the original, a gain make-up control boosts the level back up.*

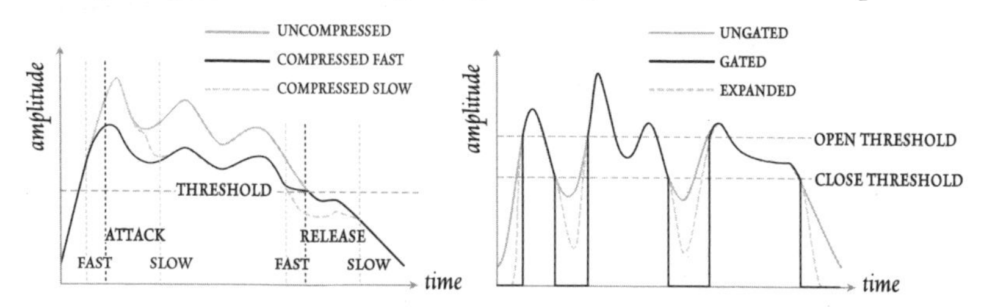

ABOVE LEFT: *Attack and release controls set how fast a unit resets once the threshold is crossed, useful for shaping transient peaks or preventing volume pumping.* RIGHT: *Noise gates often have separate threshold levels for opening and closing, letting sounds through without cutting off a quiet end.*

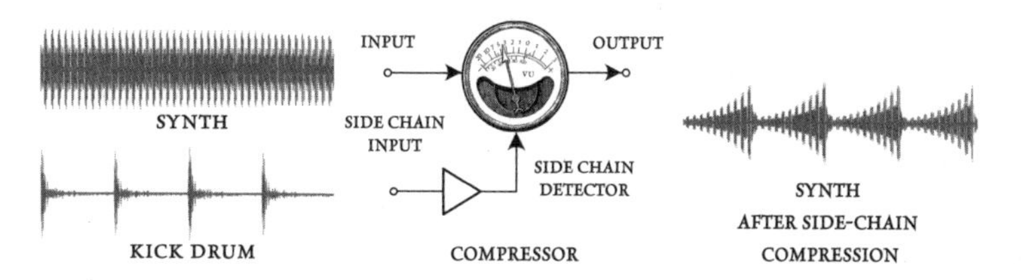

ABOVE: *Side chains. Some compressors and gates let secondary audio signals control their operation via a side-chain input. Uses include the radio* DJ *trick of automatically ducking music behind a vocal, or rhythmically dropping the level of a constant synth to let a heavy kick-drum punch through.*

DELAY & REVERB
plates, chambers and loops

ECHO (DELAY) in studios was first created using tape (*opposite below*). Modern electronic and digital delays allow precise control over delay time and decay (how quickly it fades). A short slapback echo adds excitement, longer delays emphasize phrases. Basing the delay time on the tempo can create intricate patterns with guitars, keys or drums, and combining delays with reverb and other effects can evoke etherial soundscapes. Other tricks include panning delays (which flicks the echoes between left and right), or reversing them to lead into a note.

REVERB emulates more complex acoustic reflections, such as found in a church or cave (*see page 33*). Until the 1980s, studio reverb was made by playing sound through springs, metal plates or chambers (*opposite top*). Reverbs are now mostly digital, simulated or sampled from real spaces. To simulate a forest reverb, a gun shot is recorded in a forest and then the space is recreated algorithmically (without the initial impulse) as a CONVOLUTION REVERB (*below*). A binaural recording used as a convolution sample produces 3-D reverb for headphones. Reverb can make plain vocals sound lush, or give disparate elements a coherent sonic space. A little adds a sheen, but too much can swamp a mix. To make room, some EQ or compression on the reverb helps.

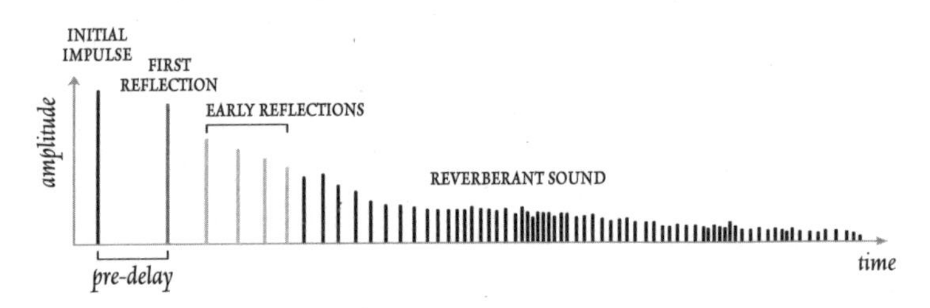

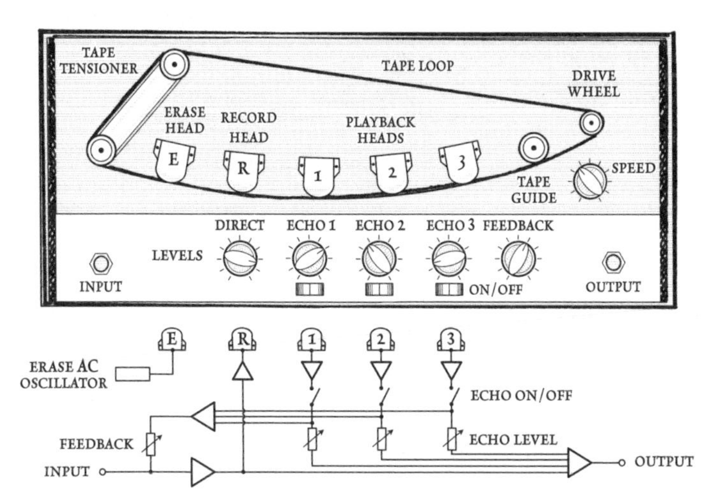

ABOVE: *A tape echo unit records sound onto a short tape loop. The tape carries the sound past the playback heads, each playing an 'echo' of the original. A feedback control adds the delayed signal back into the input to increase the number of echoes. The delay time is set by the tape speed.*

ABOVE LEFT: *a plate reverb consists of a metal sheet hung from a frame. A speaker directs sound onto the sheet, then transducers pick up the resulting vibrations to give a shimmering effect.* RIGHT: *Abbey Road's reverb chamber. Sound is played into a specially prepared room. Mics pick up the reflections from the tiled walls and hard surfaces. Moveable pillars add complexity.*

EFFECTS & ENHANCERS

it's all too beautiful

A dull sound can be made into something special by adding *effects*. They are great fun to experiment with and combine in different ways.

DISTORTION or **FUZZ** overloads the signal, clipping the top off a sinusoidal wave to make it almost square. Popular on guitar, it can also help push bass through a mix, or give an edge to vocals and other instruments.

WAH WAH is guitar classic, using a pedal to control the frequency of a low or band pass filter. **AUTO WAH** sweeps the filter frequency in proportion to the sound level - excellent for funky bass and keys.

TREMOLO is often found in country, surf and psych music. Here, a low frequency oscillator (LFO) rapidly turns the level up and down. **VIBRATO** is a little similar, using an LFO to gently wobble the pitch.

PHASING gives a swooshing, hollow effect by adding the original signal to a slightly phase-shifted copy. The resulting interference creates a comb filter with several notches. An LFO then sweeps the affected frequencies.

FLANGING makes a high flying, jet-like sound. A complex comb filter is generated by delaying the signal a very small amount (1-10 ms), then recombining with the original. The delay time is varied by an LFO, with a resonance/feedback control accentuating the robotic, metallic tone created.

Chorus and ensemble effects are similar in principle to flangers. Their slightly longer delay times (5-20ms) are wobbled by an LFO to give the impression of several instruments all playing along together.

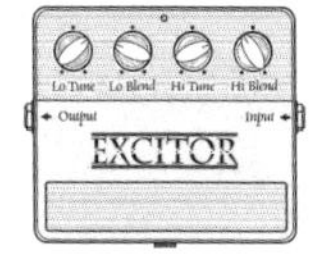

Aural enhancers add subtle harmonic distortions to give a sonic sparkle. They should be used sparingly however, as they can become wearing on the ear.

Ring modulators create out-of-this-world effects by modulating one sound onto another—great for weird resonances, warbling daleks and clashing overtones.

Vocoders also add one sound to another, but do this by separating the frequency spectrum into many slices and modulating them individually–for example, vocoding a voice onto keyboards will make a synthesizer sing.

Octavers add an extra octave or two below or above the note played. Originally achieved by analogue frequency dividers, most are now digital. They are often used to thicken up a bass or produce a really fat lead solo.

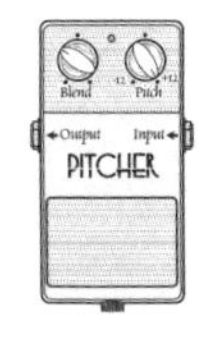

Pitch shifters digitally alter the pitch of a signal, turning a mouse into a gorilla or a baritone into a soprano. Some harmonisers can also ensure that any notes output are in the same musical scale as the input.

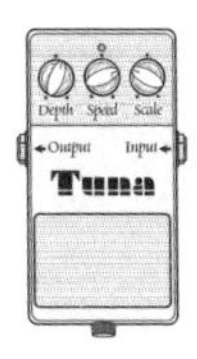

Pitch correctors and auto-tuners analyse and digitally alter any out-of-tune notes to give a perfectly in-tune output. This can vary from a tiny nudge on wayward pitches to wild yodels from note to note. Helpful on many instruments, they are a popular, if overused, vocal effect.

MIXING & MASTERING
the finishing touch

When everything has been recorded, all the various tracks are combined into a single MIX. A good mix creates clarity and separation by carefully balancing the levels, EQ, dynamics and effects of each element.

Instruments recorded onto multiple tracks, such as drums or strings may be GROUPED, allowing them to be treated as a single unit (*opposite top*). A stereo soundscape can be built by PANNING tracks to the left or right. Powerful sounds, such as lead vocals or bass instuments are often panned to centre, while backing vocals, drums, keys and guitars are spread across the stereo field to give a sense of space.

Effects like reverb or delay create ambience and depth. Rather than putting an effect on an individual track, part of the signal may be split off and sent down a BUS. A single effects processor can then act on several tracks, resulting in a more coherent sound (*opposite centre*).

AUTOMATION allows volume, EQ and other parameters to be controlled by software, whether 'riding' levels on vocals and solos, tweaking effects or removing stray pops (*below*). Mixing on different speakers can highlight odd frequencies and poor balance. Listening to the mix in a car, or quietly or in mono, can also show up faults.

When ready, the mix is BOUNCED out into a single MASTER recording. While mixing looks at multiple elements, mastering considers the whole. A mix will be heard alongside other recordings, so a mastering chain gives the mix a final polish and pumps up the volume (*lower opposite*).

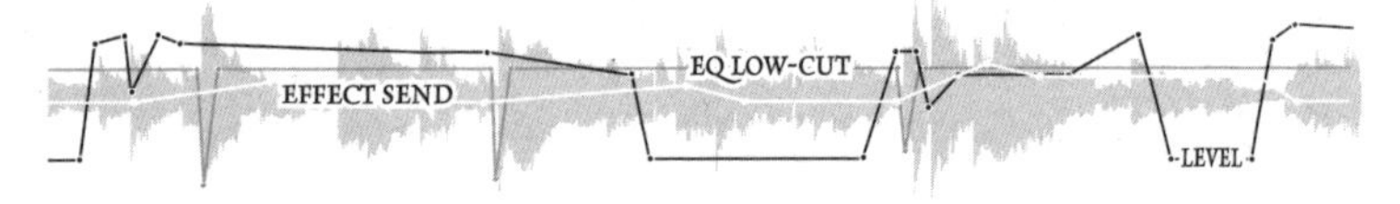

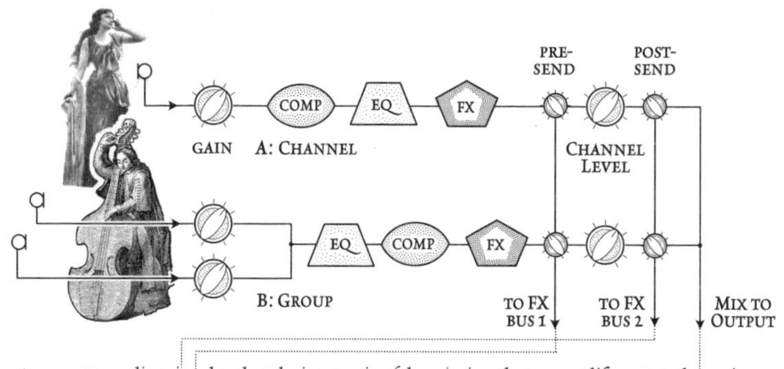

ABOVE: Recording signal paths. The input gain of the mics is set by pre-amplifiers. EQ, dynamics and effects may be applied to individual channels (A) or to combined groups (B). Part of the signal can be sent along the effects busses before (pre) or after (post) the channel volume level.

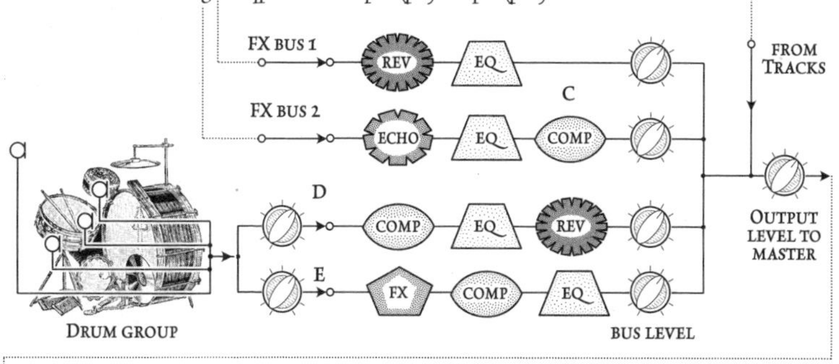

ABOVE: Effect bus reverbs and other ambient effects may be additionally compressed, EQ'd or gated to make space in a mix (C). For impact on drums, parallel compression splits the signal; one is processed normally (D), the other is effected, compressed hard and EQ'd to give punch (E).

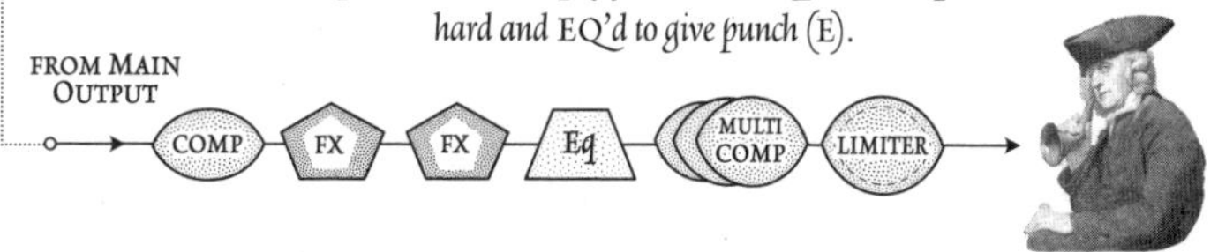

ABOVE: Mastering signal chain. A touch of compression glues the track together, whilst a hint of stereo widening and aural exciter gives a sparkle. A little EQ and multi-band compression shapes the sound and prevents pumping from stray frequency peaks, before a last squeeze by the limiter sets the level.

APPENDIX

SPEED OF SOUND IN VARIOUS MEDIA	m/s	km/s	km/h	feet/s	mph
Dry air at 20°C and 1 atm pressure	343.2	0.3432	1,235.5	1236	768
Dry air at 0°C	331.4	0.3314	1,193	1,087.3	741.3
Moist air at 20°C (90% humidity)	344.5	0.3445	1,240.2	1,130.2	770.6
Carbon dioxide at 0°C	258	0.258	928.8	846.5	577.1
Fresh water at 0°C	1403	1.403	5,050.8	4,603	3,138.4
Fresh water at 20°C	1481	1.481	5,331.6	4,858.9	3,312.9
Sea water at surface : 10°C, 35% salinity	1488.6	1.4886	5,359	4,883.9	3,329.9
Sea water 1km depth : 4°C "	1482.9	1.4829	5,338.4	4,865.2	3,317.2
Sea water 5 km depth : 2°C "	1543.7	1.5437	5,557.3	5,064.6	3,453.2
Ice at 0°C	1826	1.826	6,573.6	5,990.8	4,084.6
Iron (cast)	4994	4.994	17,978	16,385	11,171
Glass (Pyrex)	5640	5.640	20,304	18,504	12,616
Continental crust (P waves, approx)	7400	7.4	26,640	24,278	16,553
Diamond	12000	12	43,200	39,370	26,843
Solid atomic hydrogen (theoretical max.)	35,406	35.406	127,460	116,160	79,200

ABSORPTION COEFFICIENTS OF VARIOUS MATERIALS AND FINISHES

MATERIAL	125 Hz	250 Hz	500 Hz	1 kHz	2 kHz	4 kHz
Seats (upholstered)	0.6	0.74	0.88	0.96	0.93	0.85
Plasterboard (12mm)	0.29	0.1	0.06	0.05	0.04	0.04
People (adult)	0.25	0.35	0.42	0.46	0.5	0.5
Wood (decking)	0.24	0.19	0.14	0.08	0.13	0.1
Metal deck (25mm)	0.19	0.69	0.99	0.88	0.52	0.27
Glass (6mm plate)	0.18	0.06	0.04	0.03	0.02	0.02
Drapery (476 g/m²)	0.05	0.07	0.13	0.22	0.32	0.35
Brick (natural)	0.03	0.03	0.03	0.04	0.05	0.07
Carpet (Axminster)	0.01	0.02	0.06	0.15	0.25	0.45
Plaster (gypsum or lime)	0.01	0.02	0.02	0.03	0.04	0.05
Marble or concrete	0.01	0.01	0.01	0.01	0.02	0.02
Water or ice surface	0.008	0.008	0.013	0.015	0.02	0.025

MUSICAL NOTE LENGTH VS DELAY TIME AT VARIOUS TEMPOS BPM (BEATS PER MINUTE)

Note	BPM	60	70	80	90	100	110	120	130	140	150	160	170	180
1/4	time in ms:	1000	857	750	667	600	545	500	462	429	400	375	353	333
1/8		500	429	375	333	300	273	250	231	214	200	188	176	167
1/16		250	214	188	167	150	136	125	115	107	100	94	88	83
1/32		125	107	94	83	75	68	63	58	54	50	47	44	41